D0206014

On Justifying Moral Judgments

International Library of Philosophy and Scientific Method

Editor: Ted Honderich

A Catalogue of books already published in the *International Library of Philosophy and Scientific Method* will be found at the end of this volume.

On Justifying Moral Judgments

Lawrence C. Becker
Hollins College, Virginia

London
ROUTLEDGE & KEGAN PAUL
New York Humanities Press

First published 1973 by
Routledge & Kegan Paul Ltd
Broadway House, 68–74 Carter Lane,
London EC4V 5EL
Printed in Great Britain by
Alden & Mowbray Ltd
at the Alden Press, Oxford
Set in Garamond 11 point 1 point leaded

ISBN 0 7100 7524 3

To
Charlotte

CONTENTS

PREFACE

A good deal of talk about justifying moral judgments presupposes that really rigorous justifications are impossible, that moral judgments are always, at bottom, arbitrary. Moral scepticism, or as least moral agnosticism and relativism, have become something of a folkway among the literate, even while positions in current political crises get increasingly dogmatic. In short, a remarkable amount of confusion about what is possible and what is not possible in the realm of moral justification is enshrined in both philosophy and common sense. This book is an attempt to ease some of that confusion.

Since its focus is so confined, the book will inevitably lack the aura of wisdom literature. There is no opportunity to speak from the whole of one's moral concerns and convictions in a set of arguments such as this. The discussion may thus appear to be somewhat antiseptic. It may also appear to misrepresent the dynamic, dialectical nature of moral experience. Moral arguments, in practice, are seldom linear. They are more likely to be competitive exchanges than disinterested, painstaking assemblings of evidence toward conclusions. But as long as it is kept in mind that my purpose here is not to discuss moral practice directly, but rather to discuss how far one can go in reasoning about moral practice, the somewhat remorseless and static quality of the analysis will not be misleading.

As to the text itself, the first draft was complete in late 1969, and has been through a number of revisions since. I am grateful to colleagues and friends who have helped with this process. For particularly thorough critiques of the first draft, I am indebted to Alan Gewirth and F. E. Sparshott. Brand Blanshard, Charner Perry, and Mary-Barbara Zeldin also offered penetrating comments at this stage. H. L. A. Hart gave me many valuable comments on the whole of the second draft. Ronald Dworkin and Joseph Raz also read portions, and I profited from their remarks. In the final stages, Ted Honderich was responsible for calling my attention to many matters of format and style that needed improvement. These readers, of course, bear no responsibility for defects in the finished version.

Two portions of the book have been previously published: a version of Chapter II appeared in the *Journal of Value Inquiry* 6 (1972): 213–20 under the title 'Axiology, Deontology and Agent Morality: The Need for Co-ordination,' and a version of Chapter XV was published as 'Determinism as a Rhetorical Problem' in *Philosophy and Rhetoric* 4(1971): 20–8. I thank the editors involved for their permission to reuse that material here. The thesis of Chapter VII was read under the title 'Grounding Moral Judgments' at the Eastern Division Meetings of the American Philosophical Association in December, 1970.

For financial assistance, I must acknowledge the Ford Foundation and the Trustees of Hollins College for assistance with the project in its early and intermediate stages. The work was completed during my tenure as a Fellow of the National Endowment for the Humanities.

I am grateful to all my colleagues at Hollins, but particularly to those in the Department of Philosophy and Religion: Alvord Beardslee, Lamar Crosby, Allie Frazier, George Gordh, and Barbara Zeldin. My students in P & R 350 (Spring 1970) were also of great assistance. In particular, I am indebted to Katherine Grieb and Dianna Strange for useful suggestions. Shirley Henn, reference librarian of the Fishburn Library, Hollins College, gave special bibliographical help.

Finally, to my wife, who helped more, and in more ways, than anyone else, I can only offer the dedication.

Hollins College, Virginia L.C.B.

I

INTRODUCTION

Philosophic talk about morality goes on endlessly, and because it is so largely repetitious and so largely ineffectual, it seems, at moments of weariness, to be pointless as well. The persistent plausibility of moral scepticism is no comfort; it reinforces every discontent with philosophy.

The purpose of this book is to destroy the plausibility of moral scepticism, and by doing so, to blunt some of the dismay any man must feel in observing the frequently fruitless encounter between moral philosophy and moral problems. The plan here is not to develop a substantive moral theory, though inevitably some substantive principles will be recommended or ruled out along the way. The plan is rather to lay bare *a set of procedures for the justification of moral judgments* which is both straightforward enough and thoroughgoing enough to win general acceptance as an *adequate* set of procedures.

The emphasis on the justification problem is not new. Indeed, there is a whole school of recent moral philosophy preoccupied with it: the so-called 'good reasons' approach to ethics. And it is clear why such preoccupation is important. Scepticism stands or falls with the outcome. If one cannot find a way to get thoroughgoing reasoned justification for *any* moral judgments, the sceptic's thesis is the only consistent extension of the evidence. At the very least, avoiding it will be a constant struggle. If, on the other hand, one can find a way to justify at least some moral judgments, scepticism is disabled, and the claim that moral philosophy is useless in moral affairs is demolished.

Previous attempts to lay out justification procedures have run into heavy criticism, and it is safe to say that none has won general acceptance as an adequate set of procedures. Some are said to conceal logical errors (e.g., Marcus Singer's attempt to ground moral judgments in the 'Generalization Argument'). Others are attacked for merely pushing back the point at which arbitrary assumptions must be made (e.g., Hare's analysis of moral arguments, Searle's proposal for getting from 'is' to 'ought'). Still others are suspected of disguising either circularity or arbitrariness in their appeal to the criterion of 'reasonableness' as a fundamental limiting principle (e.g., Toulmin and perhaps Baier).

No extended evaluation of these arguments and counterarguments will be undertaken here. Indebtedness to these controversies will be obvious in almost all of the chapters to follow. But the first debt which must be acknowledged concerns how to organize and display a justification procedure. For in the course of the controversies raised by previous attempts, it has become clear that if one is to succeed in getting an adequate set of procedures—and more than that, in winning general agreement that the procedures are adequate—he will do well to stick exclusively and explicitly to questions of procedure. Further, he will have to organize his arguments so as to bring out forcefully (1) the way in which the procedures settle the grounding problem; (2) the way in which they settle or avoid certain longstanding philosophical puzzles about morality; and (3) the degree to which they diminish the ambiguity of moral judgments. A few remarks on each of these matters will serve as an introduction to the main line of argument.

1 THE PROCEDURE PROBLEM

The insistent focus on matters of procedure in what follows will no doubt be an irritation to those impatient to get on to weightier matters. The irritation is unfortunate, but these arguments can succeed in their purpose only by sticking rather single-mindedly to their point, which is to remove a major impediment to substantive work in moral philosophy. If they were themselves to shock or surprise or excite (on substantive, rather than procedural, issues) they would likely raise more problems than they settle, and thus not serve their purpose at all.

But even so, one may wonder whether it is reasonable to devote an entire book to procedure. Granting that the problem is important, and that many mistakes in philosophy (as well as many failures to make others see that an argument is valid) are due to abbreviation, are procedural matters important enough to require such extended treatment? In brief, yes. But in addition, the focus has some special values, both with respect to moral practice and moral theory.

With respect to moral practice, it may be remarked that a treatise on procedure is likely to be of more enduring use than an attempt at a 'system' of moral philosophy. Treatises which concentrate on building systems of principles are usually the least use in practice. For even if one agrees with the principles wholeheartedly (and that is unusual; usually one needs to re-state them, and thus 're-derive' them), it must still be decided whether any of the principles apply in a given case, and if so, how. Thus one is faced with the need to know procedures after all. And because the principles most moral philosophers arrive at are rather unremarkable when all is said and done—that is, are often startlingly similar to common sense views or cultural mores—the time spent in learning a system can seem a waste.

As a rough piece of moral psychology, it might also be remarked that if one is to use moral principles and judgments adequately as a basis for action, he needs to 'make them his own.' While this can of course be done by 'internalizing' the pronouncements of some authority (e.g., the moral-system builder), it can also be done, and perhaps better done, by thinking things out with somewhat less overt guidance than that given by a 'system'. Considerations such as these alone recommend a focus on procedure.

With respect to moral theory, the arguments of this book will show that the procedure-focus can result in significant theoretic simplifications, avoid some perennial impasses, and secure at least a few agreements in otherwise unlikely places. So again, concentration on the given problem is valuable.

But fundamentally, the warrant for exclusive focus on procedure rests with its usefulness in disabling moral scepticism. And that use requires successful completion of at least three tasks. To wit:

2 THE GROUNDING PROBLEM

It is an unassailable truism that one cannot prove everything. And in the absence of some 'self-evident truths' for morality, or some other philosophically acceptable grounding for justifications, every attempt at a reasoned justification for moral judgments seems destined to end either in an indefinite regress of reason-giving, a logical circle, or some arbitrary assumptions. It is hard to see any other alternatives. The grounding problem is thus the problem of putting a non-arbitrary stop to the process of reason-giving in the justification of moral judgments.

There have been attempts to ground justifications with intuitions about values, intuitions about duties, natural norms, natural laws, and even 'morally transparent cases.' But all of these attempts are objectionable: intuitionism, moral sense theories, and the 'transparency' theory because they have provided no empirical way of distinguishing their results from the articulation of deep (but learned) feelings, convictions, and dispositions; natural norm and natural law theories because they have provided no convincing assurance that the norms or laws are anything more than (logically) optional human contrivances—at least in terms of their use as a guide to reasoning out what we ought to do.

If an adequate set of justification procedures is to be found, however, the grounding problem must be solved. If it cannot be solved, justification procedures will never be adequate in the sense required for the refutation of scepticism. I will argue in Chapters VII, X, and again in XIX that there is an unobjectionable procedure for relying on natural norms to ground moral judgments.

3 PHILOSOPHIC QUAGMIRES

In addition to the grounding problem, one can cite a long list of philosophic controversies which seem to block justification procedures at one point or another. Are the fundamental principles of morality to be expressed in terms of good (i.e., the best or most valuable) conduct? Or are they to be expressed in terms of moral duty, or perhaps moral character? What is one to make of the supposed distinction between values and facts? Are values commensurable and additive? Can one derive an 'ought' from an 'is'? Can punishment be justified on utilitarian grounds alone?

The list is a long one, and put so baldly, seems a pretentious catalogue. But each of these controversies (as well as others) stands in the way of some part of an adequate set of justification procedures. And so each will have to be dealt with in turn—not, necessarily, by providing a solution to the problem itself, but at least by providing a way of circumventing it, together with adequate assurance that such circumvention is unobjectionable. Much of the argument of this book, then, will deal with steering clear of traditional controversies.

4 THE AMBIGUITY OF MORAL JUDGMENT

What cannot be steered clear of are some hard facts about the uncertainty of moral judgment. Insofar as the consequences of one's acts are indeterminate and/or indeterminable, any rational assessment of conduct-options will be incomplete. This incompleteness plagues retrospective judgments as well as prospective ones. A case can hardly ever (in principle) be closed against the possibility of new evidence. That is why moral arguments contain so many *ceteris paribus* clauses, and why moral argument is often described as 'open-textured,' 'loose jointed,' and the like.

Some writers (e.g., Sartre) have made much of this ambiguity, leaving the impression (whether intentionally or not) that moral acts are at bottom all equally arbitrary and that the search for rational assurance of the justifiability of conduct can only be a false and dangerous course—an attempt to escape the reality of our simultaneous ignorance about, and responsibility for, our acts.

Such a conclusion would naturally be fatal to the enterprise of this book, and so its challenge must be met. It will not be argued that the procedures proposed here are decision procedures; but it will be argued that they are occasionally adequate to yield decisions. And that much, even alongside the notable 'looseness' of moral argument, will be enough to blunt the sceptical force of the ambiguity issue.

5 AN OVERVIEW

The plan of the book is to begin with an argument for the need to co-ordinate three 'types' of moral theory with a comprehensive set of justification procedures. Those procedures will then be displayed, beginning with an analysis of valuation, evaluation, and the grounding of value judgments, continuing through the justification of ought-statements and the ascription of responsibility, and concluding with some remarks on 'agent morality'. Along the way the problem of distinguishing moral contexts from non-moral ones and the question of why one ought to be moral at all will also be considered.

For the reasons previously mentioned, the focus will be exclusively on justification procedure. Two further items about this need to be noted. First, the procedures outlined here are designed to answer the question: *What would one have to do to get thoroughgoing reasoned justifications for moral judgments?* The procedures are not put forward as a description of how people do in fact go about justifying their moral judgments, though that is precisely where the analysis will often start—with a survey of how it is done in practice. Nor are these procedures put forward as a recommendation for how people ought to behave in concrete moral situations, though some of the procedures are surely applicable to moral practice. The procedures are rather a description of how a moral *theorist* can assemble justifications for moral judgments. Thus the 'point of view' here (i.e., the 'person' doing the justifying) is that of the disinterested philosopher—operating, to be sure, on real (as opposed to purely 'academic') problems, but not confined by the demand for a decision at a specified time, and not permitted to introduce *ad hoc* arguments on the basis of fortuitous and special circumstances. (For example, on the basis of the fact that 'men generally agree that. . ., so we can move on to the real source of controversy.')

Second, it must be remembered that since the points here are procedural ones, the examples chosen to illustrate the procedures are in the main *merely* illustrative. Haggling over the appropriateness of a given example is not likely to constitute an objection to the main argument. This is an ordinary enough point, but one which ought especially to be attended to during the discussions of 'presumptive criteria' (Chapters VII, X and XIX).

Without further preliminaries, then, I shall turn to the question of justification procedures for moral judgments.

II

AXIOLOGY, DEONTOLOGY AND AGENT MORALITY

It has become something of a tradition for ethical theorists to organize their efforts around one of (roughly) three sets of concepts: value, obligation, or moral character-traits. The result is a three-fold distinction in types of moral theory: (1) axiological theory, which takes concepts of value to be primary (if not primitive) and ultimately decides (if not redefines) questions of duty and virtue in terms of value questions; (2) deontological theory, which takes concepts of obligation and duty to be primary (if not primitive) and ultimately decides (if not redefines) questions of value and virtue in terms of duty questions; and (3) moral-agent theory, which takes concepts of virtue and vice as primary (if not primitive) and ultimately decides (if not redefines) questions of value and duty in terms of questions about the good or virtuous man. These types of moral theory are quite distinct in tone, emphasis, and sometimes substance. And there are advocates for each who find the others quite distasteful—even beside the point.[1]

Each type of theory can make a plausible case for its usefulness, but each is also subject to serious difficulties. The inter-scholastic jostling for primacy among them has been illuminating, but it has not secured agreement on the superiority of any one of the three theories, and the prospect of perpetual conflict on these matters is not pleasant—at least from the standpoint of an attempt to discuss justification procedures.

The contention here will be that these three types of moral theory need to be co-ordinated, and that this co-ordination can be usefully accomplished in terms of a comprehensive set of justification procedures which has none of the partisan overtones of exclusively axiological, deontological, or agent-morality theories. This chapter will argue the need for co-ordination—that is, reasons for thinking that any attempt to make any two of these types of theory thoroughly subordinate to a third will be unsuccessful. Subsequent chapters will present the justification procedures in terms of which they can be co-ordinated.

I SOME ORDINARY ASSUMPTIONS

The following assumptions, while not unproblematic, are quite commonly made in moral philosophy, and will be used here without serious elaboration or defense.

(1) It will be assumed that *existing elements* of what one loosely calls 'moral life' or 'moral experience'—e.g., valuation, evaluation, the practices and institutions associated with making, keeping, and enforcing obligations—constitute an inescapable part of the data to be dealt with by moral philosophy. While no assumption is made about the ultimate justifiability of a given practice merely from the fact that it exists, a moral theory which rules it out (either with an explicit injunction against it or by simply not 'having a place' for it) may not do so arbitrarily without being considered, to that extent, objectionable. This is so because philosophical theories are (by definition) attempts to get reasoned understandings of things. When philosophy lapses into arbitrariness, it fails to meet the conditions which define it as philosophy. Quite generally, then, regardless of whether a moral theory accepts or rejects the existing intuitions, practices, and institutions of moral experience, it may not properly do either without reasons. When a theory does either in an arbitrary way, one is justified in judging it objectionable in direct proportion to the importance of the issue (the 'given') at stake. A significant part of the moral philosopher's activity, then, is in dealing with the given, or existing, elements of moral experience.

(2) It will be assumed that a part of the moral philosopher's task, in dealing with the aforementioned data, is to articulate as fully as he can (in propositions) and present in reasoned arguments just what those data are, and what their consequences are, and have been, and are likely to be. In other words, a part of the

philosopher's task is to express propositionally (and arrange theoretically) both what men do in the way of valuing, evaluating, etc., and what they believe about what they do—what they 'know' in the sense of having experienced, what they 'intuit,' what they 'would want to say,' what the 'common sense' of the race is.

(3) It will be assumed that the point of doing this is not to raise 'common sense' or 'what we would want to say' to the level of the ultimate arbiter of validity for moral theory. The point is simply to deal unprejudicially with what is given—namely the totality of practice marked off as 'moral' and that totality of experience, conviction, conclusion, and belief similarly marked. Dealing with the given in an unprejudiced way merely means neither accepting it (as 'true' or as an arbiter of theory), nor rejecting it, without reasons.

(4) It will be assumed that when one is trying to 'get things into propositions' (which is, of course, even more strenuous a task than 'getting things into words') he can and often does misrepresent those things. Precisely how a proposition can misrepresent something is of course a difficult matter, but it can reasonably be avoided here.

(5) It will be assumed that one may legitimately argue that a particular attempt to articulate something misrepresents what it claims to represent by showing that—for some reason—'we wouldn't want to say that,' or that the attempt in question has consequences which, when stated, are not 'what we want to say.' The point is that when a proposition misrepresents the state of affairs it purports to represent—by virtue of an unfortunate emphasis, a misleading focus, or by some more straightforward type of error—one would expect a protest of precisely this form: 'That isn't quite what should be said,' or 'No, I wouldn't want to put it that way.' The mere existence of such objections does not, of course, mean that there has in fact been misrepresentation. But the persistence of the objections, together with some explanation of just where the proposition in question is misrepresentative, surely builds a reasoned case against the adequacy of the proposition.

It may bear reiteration, however, for those philosophers irritated by arguments based on 'what one would want to say,' that the point here is not that what we want to say has any special sort of *authority*. The point is simply that the philosopher is committed by the nature of his enterprise, to giving a reasoned defense of his views. An 'objection' from common sense, or in terms of 'what we would want to say,' requires a non-arbitrary answer. To the degree that the philosopher can give no reasons for deviating from 'common sense,' then to that degree his deviation is objectionable as a philosophical position. Of course it also bears mentioning that a philosopher must defend any *adherence* to common sense or 'what we would want to say.' Pointing out that a position deviates from the typical does not by itself constitute a philosophic objection to it. One needs additional argument for that: e.g., of the sort J. L. Austin often made.[2]

All of the assumptions just enumerated are, as noted above, somewhat problematic in themselves. But they are not extraordinary for discussions in moral philosophy—although they seem to be used implicitly more often than explicitly. When one or another of the arguments to follow seems to require a more careful explication and/or defense of one of the assumptions, it will be done in the process of presenting the argument.

2 THE THREE THEORIES: STRENGTHS AND WEAKNESSES

When one examines the various types of ethical theories mentioned at the outset, he is struck by two things: first that there is a significant region of common moral experience which each of the types seems better suited than the others to 'express'; and second, that there is another region of common moral experience which each seems inadequate to 'express.' The most prominent example (though by no means the only one; punishment is another) concerns the concept of duty. There are at least two areas here in which the deontologist has to struggle to make his views plausible due to his insistence that duty is the fundamental moral category. And there are equally important areas in which axiological and agent-morality accounts are objectionable.

Consider the problem of justifying propositions which assert or deny the existence of duty. Axiologists attack such a problem by way of an analysis of the *values* involved: the values of the action required by one's duty: and the values of requiring such actions of anyone. One justifies a duty, then, by reference to whether it can be considered 'right conduct' (i.e., more good than bad).

Agent moralists attack the problem by way of an analysis of the compatibility of the duty in question with what it means to be a good, virtuous or just man. They justify a duty by reference to whether it can be considered virtuous conduct—conduct 'in moral character.'

Now one must acknowledge the importance and reasonableness of both sorts of argument on the question of the justifiability of one's duties. Yet the deontologist, insofar as he insists on using the concept of duty as *the* fundamental of his moral theory, cannot accept the *need* for such justification, and is in the awkward position of affirming that one may have duties which are unjustifiable—both in terms of the values they embody or produce, and in terms of the moral character their performance exemplifies. This is an awkward position for the deontologist in the sense that one finds it strange (unusual and vaguely dangerous) to insist that 'None the less (i.e., in spite of there being no good to be gained) it is one's duty to do *x*.' Or worse, that 'None the less (i.e., in spite of the demonstrable preponderance of evil involved) it is one's duty to do *x*.' Yet the discomfort one feels with such assertions is inconsolable within a deontological scheme due to the deontologist's explicit ruling out of any means of consolation: that is, of any means of legitimizing a duty or obligation in terms of something equally fundamental.

'Supererogative' situations are another place where the deontologist must struggle to make his views plausible, while axiologists, and particularly agent-moralists, have no problems. For while 'going beyond the call of duty' is good and right (as the axiological account would have it), and while it is an admirable character trait (as agent morality would have it), there is something odd about expressing approval of it in deontological language. 'Duties to go beyond one's duties' has the hint of a contradiction in it—and rather misrepresents the character of the conduct referred to, most would say. This cannot help but be an uncomfortable moment for the deontologist.

Now, what is to be done with such discomfort? If it indicates that deontological theories misrepresent, or at least conflict with, certain portions of 'common sense,' how is one to assess the overall adequacy of those theories? It is not, of course, necessary to bring a theory into agreement with 'What we would want to say' at every point. But the assumption is that some reasons must be offered for allowing such a conflict to stand, and that if no such reasons can be offered, the theory—and not 'common sense' —is to that degree objectionable. It, and not common sense, is objectionable because it, and not common sense, is the self-avowed attempt to get a reasoned understanding of things. Deontologists must concede, ultimately, that on the matter of unjustifiable duties and the matter of supererogation, their theories are objectionable in this way.

To say that a theory is objectionable, of course, is not to say that it is untenable. Deontologists will argue that the advantages of the deontological approach make a bit of discomfort worthwhile. And it is surely true that for every case which can disquiet a deontologist, there is a matching embarrassment for other theorists. It is in just those cases that the advantages of organizing one's moral theory around the notion of 'the "ought" of obligation' are most obvious.

To be more specific, one thing that is embarrassing about going from a deontological to an axiological or agent-morality account of duty and obligation is the way in which the focus of concern gets shifted from the 'binding' or imperatival force of the ought statement to other things: to the value of being bound, or even simply the value of the act in question; or to the issue of what it means to be a good man. There are cases in which such a shift bothers no one but a hypersensitive deontologist. But there are other cases in which, no matter what his predispositions, one must admit that these shifts in emphasis somehow misrepresent 'what we want to say' about the case. Sometimes, what we want to emphasize is precisely the way in which we are 'bound' to do x, and remarks about the good to be achieved seem to dilute that emphasis. They dilute it by calling attention to the way in which a man might decide whether or not to do x without primary reference to duty or obligation at all, but rather with primary reference to the way in which doing x (or even being bound to do x) is, on balance, good. And if one grants any legitimacy at all to the practice of organizing some regions of life in terms of obligations (or at least responsibilities), he must grant that some of the point of doing so is lost by this dilution. On the matter of expressing the 'binding' nature of obligations, then, I take it that deontological theories excel, and axiological and 'agent morality' theories are objectionable.

Objections against both axiology and deontology, however, can be lodged on precisely the grounds on which the battles described above were fought: that they both conflict with another portion of our moral experience. It can be argued with considerable force that, at least in some cases, a good deal of what we want to say about the morality of human conduct is grossly distorted both by haggling over the balance of goods and bads on the one hand, and by the messy business of assigning responsibility, considering excuses, and deciding whether and if so how to intervene in our fellows' affairs on the other. There is at times a certain loss of nobility in such practices—a loss we often want to express by insisting on shifting the focus of discussion to the 'sort' of man we are confronted with, whether a good or a bad man (never mind temporary aberrations), whether virtuous or vicious. What our experience suggests is that we—and hence by extension, our fellows—are also capable of the dignity and heroism of conduct which cannot be recommended in terms of what can be known of its value, nor demanded as a consequence of one's obligations. And that we are equally as capable of the sorts of horrors for which all the adverse value descriptions and all the 'deontic rebukes' imaginable are inadequate. In such cases we tend to rely on character descriptions (if, indeed, we articulate at all), and talk of the types of conduct involved. And in such cases 'agent morality' seems best suited to express what we want to say, while both axiology and deontology are objectionable. They are objectionable, again, because what they force us to say conflicts with what we want to say, and because they cannot resolve these conflicts in an unarbitrary way.

Agent morality, of course, has some significant weaknesses also. For one thing, we are all well aware of the fact that a noble man —acting fully in character—can do insupportable wrongs. And it is no surprise that classical Greek literature devoted itself so often to this theme. For the Greeks were prone to use a virtues–vices approach to morality, and the 'noble immorality' is a poignant, puzzling case for such an approach. In addition, of course, one cannot help but feel that the 'morality of aspiration,' as it has been called,[3] is often too vague an instrument to use in adjudicating the messy details of day-to-day human interactions. Character assessments are usually beside the point in small claims court.

3 THE UNELIMINABILITY OF WEAKNESSES

So each of the major types of ethical theory under discussion is objectionable at some point. The task now is to speculate a bit on the explanation for and the implications of this situation.

As has been implied before, one must be careful to observe that a theory's being objectionable in this way does not mean that any of its propositions are untrue, or even that the theory as a whole is untenable. One is unlikely to get any theory, ever, which is completely unobjectionable. The situation with axiology, deontology, and agent morality seems to be that each can find enough flaws in its opponents to offset the impact of its own difficulties. Each, thus, remains a 'defensible' theory. What I want to consider is whether this deadlock might not be due to the limitations of the theoretic enterprise itself. Specifically, I want to consider whether the problem might not be in the constraints of purely 'propositional' expression.

Whenever a theoretic enterprise is forced to come to terms with (either by giving expression to, or giving an 'account' of) the full complexity of a region of human experience, it faces peculiar difficulties. Not all philosophic theories are faced with such a task, but moral philosophy is. At least, that assumption is commonly made, and was explicitly noted here. There are good reasons for the assumption. Moral philosophy arises, one supposes, as a response to needs generated in the situations marked off as 'the moral life' or moral experience, and when it fails to meet those needs in some way—either by positively misrepresenting its data, or by refusing to give reasons for not accommodating itself to them—it is flawed with respect to the demands of its original reason-for-being.

But the task of getting the full complexity of a region of human experience 'into' an array of propositions is a formidable one at best, and one should not be surprised if it were demonstrably impossible. Two sorts of considerations lie behind this (admittedly vague) assertion of the limits of language.

The first has to do with one of the things noted in a particularly pointed way by Michael Polanyi:[4] that we literally 'know more than we can say.' For example, every human experience is capable of an indefinite number of true (partial) descriptions, which would be immediately recognized as true by one who had the experience. There is, thus, a perfectly literal sense in which we can never express all we 'know' about our experience. Any attempt to express 'all' of what we know in this sense—rigorously theoretic or not—will be vulnerable to the objection that one has failed to mention something or other.

This situation by itself is not devastating to a theorist, for he may argue, by analogy, that just as it is possible to say things that are true of all real numbers without being able to enumerate them, so it is possible to say things that are true of the full complexity of a moral experience without being able to enumerate its components. The task, a theorist might argue, is (simply!) to find expressions which will 'apply' to the 'full' experience.

Metaphorical and analogical problems aside, the theorist's rejoinder ignores one additional constriction of a theoretic array of propositions. Propositions—at least in so far as they have (in a grammarian's sense) a subject-predicate structure—select isolable items for one's attention. The familiar sorts of propositions (when attended to and understood, of course) function so as to mark off for, and direct attention to, selected items—whether those items are objects in the usual sense, or qualities, or functions, or activities is no matter. The effect is to focus attention and concern on some items rather than others. And this is precisely the point at which the shoe begins to pinch, for it is precisely the resultant misemphases, or misleading, or otherwise objectionable foci that each of the moral theories discussed here objects to in its rivals.

Furthermore, the arrangement of groups of propositions in theoretical arrays is inevitably a perpetuation of the 'directedness' —and therefore of any misemphasis—of the fundamental propositions of the theory, in so far as it insists on understanding (justifying or even defining) the subordinate ones in terms of those fundamental ones. The more rigorously one theorizes axiologically (i.e., the more he insists on actually *defining* all other ethical terms in terms of 'good' and 'bad,' or the more he insists, in some other way, on subordinating all other moral judgments to judgments of value), the more unsatisfactory his references to duties will be. For the focus on values, as was argued earlier, slights the notion of 'being bound' to a course of conduct. Similarly for deontology and agent morality: the more rigorously theoretic they are, the more unsatisfactory will be some of their (subordinate) references.

When one combines these (admittedly vague) remarks about the nature of theorizing with the observation that the weaknesses of each type of theory are really built into its nature, the conclusion can only be that the weaknesses are uneliminable, and that their effect is unavoidable in any hierarchical scheme which subordinates two of the types to a third. The supererogation problem, and the unjustifiable duties problem, are built-in to the business of explicating the whole of moral experience in deontic language. The dilution of the 'binding' nature of some situations is inevitable for axiological accounts. And the noble immorality, as well as ineffectiveness with the messy details of day-to-day existence, are intrinsic to agent morality.

4 THE NEED FOR CO-ORDINATION

What these reflections suggest, then, is that any rigorously theoretical (that is to say, philosophical) ethic of one of the types discussed here will be insupportably naive until it consigns itself to service as one among several co-ordinate theoretic efforts. And these reflections further suggest that effort is ill spent in trying to subordinate the best consequences of one of these theoretical formulations to another. Effort is much better spent in trying to co-ordinate various theoretical results and in investigating the way in which theoretical results in general enter into practical affairs (e.g., enter into decision procedures in the moral life). For if the inadequacies of axiology, deontology, and agent morality are in fact due to the sort of limitations indicated, then their intercollegiate rivalries are pointless, and the truly pointed problem is how their co-ordination can most adequately respond to the needs which continually give rise to moral philosophy.

NOTES

1 For a strong argument against 'law-type ethics,' see G. E. M. Anscombe, 'Modern Moral Philosophy,' *Philosophy*, 33 (1958), pp. 1–19.

2 See, for example, J. L. Austin's 'A Plea for Excuses,' in his *Philosophical Papers* (Oxford: Clarendon Press, 1955).

3 See Lon Fuller, *The Morality of Law* (New Haven: Yale University Press, 1964).

4 Michael Polanyi, *The Tacit Dimension* (Garden City: Doubleday, 1968).

III

VALUES AND JUSTIFICATION PROCEDURE

A likely place to begin an inquiry into justification procedures is with the general theory of value. Attempts to justify moral judgments inevitably face questions about values, no matter whether those judgments concern right, wrong, duty, obligation or moral character. The value-questions they face are not only those about the appropriateness of structuring a justification solely in terms of a 'calculus' of values, but are also those logically prior questions about the nature of value itself.

But these matters are among the most tangled controversies in philosophy, and it is clear that much care is required in formulating the questions which begin an inquiry into them. To ask 'What is value itself?' is to suggest (and not too covertly) that one is looking for a *thing*—an entity or quality. And readers need not be reminded of the scarce rewards of that quest. The way in which an inquiry is undertaken has consequences for what phenomena one is most likely to 'see,' the phenomena one 'sees' have a good deal to do with the conclusions he draws, and the conclusions he draws (as they collide with further considerations) determine where and how he will have to engage in subtleties—qualifications, *ad hoc* explanations, insistence on a special emphasis, and the rest.

The starting point of this study need not be general theory of value. That is, there is no *a priori* reason to suppose that a study of justification procedure *must* start there. One might instead begin with an analysis of duty or obligation, or perhaps with virtue, or the notion of a good man. All these items are relevant to an inquiry into justification procedures and all will, in due course, be treated, not as subordinate concerns of a grand axiological scheme, but as features co-ordinate with questions of value. But one begins where it seems most convenient—where he has found (or suspects) the most return for the least effort, the clearest progress from one issue to another, and the most inclusive account of moral problems. Questions of value seem, in those terms, the best candidates for a starting place.

1 THE APPROACH TO BE USED

The usual approach (since Moore) has been to go directly to the question of meaning—either the meaning of 'ethical terms' ('good,' 'bad,' 'right,' 'wrong,' etc.) or the meaning of whole utterances in value discourse. For it is not even clear, at the outset, whether value utterances are 'judgments' in any straightforward usage of the term (i.e., statements; sentences that can be true or false). And it would be foolish to run on about justifications before one knows what sort of thing he is supposed to justify. But it will be shown by what follows that considerable simplification and clarity is gained by reframing the inquiry, and initially avoiding questions concerning meaning. Here the analysis will be organized to investigate first what sorts of things people are doing when philosophers would say they are 'placing a value' on something, 'determining' a thing's value, 'assessing' the value, and so on. In short, what is proposed is a direct approach to description of the activities philosophers would be prepared to call *valuations*.

This is not done to achieve novelty. Indeed, what is new in this account is only the arrangement of some very standard observations about valuation, and thus—by that fact—the inferences which the observations encourage. (It should be no surprise that different arrangements encourage different inferences, and that some arrangements of evidence—like Lewis Carroll's exercises with sorites—discourage making any inferences whatsoever.) Further, the order in which conclusions are reached in what follows minimizes the number of subtleties which need to be introduced to meet objections.

2 SOME CAUTIONS

Put as a question, then, the inquiry into value questions will begin by asking, 'What are people doing when they value things?' Several cautionary remarks are in order, however. (1) 'Things' is taken in the broadest possible sense—in the sense used in words like 'something,' 'anything,' or 'nothing' when the discourse is in no way restricted to 'substances' or 'physical objects.' Included, then, are actions, persons, personalities, institutions. . . in short, anything whose word can (grammatically) stand in the blank of the schema '—is good.' (2) 'What' should not be allowed to conceal any assumption about the number of activities involved. When one says, 'What are they doing?' there is perhaps a vague hint of the sense, 'What *one* thing are they doing?' No such restrictiveness is intended in the question 'What are people doing when they value things?' There may be a plurality of things which go under that name and which have only the name in common. (3) 'When they value things' should not, at this point, suggest anything about whether people know, find, discover, perceive, sense, or intuit values, or whether they create, assign, project, express, or invent them. That is, a decision on the question of whether things 'have' values or whether people 'give' them values does not lie behind the question, 'What are people doing when they value things?' (4) There is one matter of meaning involved here. The questions, 'What are people doing when they value things?' and 'What does it mean to say people value things?' are treated as interchangeable. The way to find out what valuing amounts to is (oversimply) to find out what activities people are prepared to call valuational and then to describe the features of such activities. This is to ask for the meaning of some expressions, though the question of meaning here is not the question of the meaning of utterances *in* value discourse, but the meaning of the assertion that 'x is valuing y.'

3 CONDITIONS TO BE MET BY AN ACCOUNT OF VALUATION

Theorizing about the nature of value starts from common ground:

the practices, claims, hopes, insistences, and demands made by people engaged in the business of valuing. People appraise, evaluate, 'figure out,' determine, prize, appreciate. . . all of which *prima facie* come under the heading of 'valuing.' And claims are entered for those acts, hopes held, and so on which any adequate theory of value must come to terms with. Needless to say, a theory cannot be bound, *a priori*, to accept all the claims and uses, to satisfy all the hopes. This common ground is, after all, at least superficially confused and contradictory: e.g., 'There's no disputing taste.' 'I really can't decide the worth of this painting; I'm not an expert.' But the theory will have to *account* for all the elements of the common ground, and to do so without prejudice it must start with them all on 'equal feet'—i.e., treated as equally 'valid' until there are reasons to the contrary.[1] So the general theory of value will have to meet at least the following conditions:

THE PROBLEM OF CLASSIFICATION

Any description of an act of valuing will have to have some warrant—that is, there will have to be some unambiguous reason for deciding that what is being described is being properly described, and being properly dubbed a valuation. Suppose someone calls cracking an egg a valuational act? Must I accept it? Must I add it to the lexicon without further inquiry? Or may I ask for an explanation? This is a delicate matter, for to demand an explanation of how cracking an egg is a valuation seems to be to require that some 'essential' similarity be shown between that act and other acts agreed to be valuational. That is, it seems to suppose that all valuation is essentially one kind of act.

Yet this is precisely a conclusion one does not want to prejudice. One wants to keep open the possibility that there is more than one type of valuation. The problem, of course, is to say just how several 'essentially distinct' activities can reasonably be given the same title (valuations), while some other activity—also distinct from the rest—can reasonably be excluded. The problem is, thus, the standard one of trying to decide 'What sorts of things are properly described as—?' where the task is to fill in the blank without, on the one hand, engaging in purely arbitrary stipulations, or, on the other, being forced to fill it with anything some perverse respondent might suggest. Such decisions can only be defended, it seems, by the use of some rough and rather imprecise rule for family resemblances, or for 'what we would say.' If one can find no reason at all to suppose egg-cracking to resemble other valuational acts, or no reason to suppose people would unperversely call it valuational, then one judges it not to be a valuation. In any case, the limitations and uncertainties of this sort of classification should by this point in the century be obvious, as should the fact that suspicion must follow any claim to have found an exhaustive list of types of valuational acts, or an exhaustive real-definition of what it is to value something. More about this difficulty later.

THE RELEVANCE OF EVIDENCE

The descriptive account of valuation will have to account for or dispose of the persistent contention that the properties of things are somehow relevant to the question of their values. People persist in citing properties (e.g., strength, color) in support of value claims; the theorist is obliged to make something of this. Whether he makes sense or nonsense of it is another matter.

MISTAKES

The description of valuation will have to account for or dispose of the contention that it is sometimes possible to be mistaken about a thing's value. Moral and aesthetic arguments persist, and they are not always thought to be in vain.

RELATIVE VALUE

The description of valuation will have to account for or dispose of the contentions relating to whether or not a thing's value can change with its circumstances. If this is occasionally the case, is it always the case?

DE GUSTIBUS NON DISPUTANDEM EST

The description will have to account for or dispose of the contention that—at least in some cases—there is no disputing taste.

THE COGNITIVE STATUS OF VALUATION

The description will have to account for or dispose of the contention that valuation cannot ever be a purely cognitive act. 'It is characteristic of living mind to be *for* some things and *against* others. This polarity is not reducible to that between "yes" and "no" in the logical or purely cognitive sense, because one can say "yes" with reluctance or be glad to say "no." '[2]

This is by no means an exhaustive list of things a philosophically adequate account of valuation must do, only a suggestion of the range of difficulties it must meet. It must meet these difficulties simply because they arise from practice—from the uses people have traditionally made of value judgments. The uses may be wholly or in part mistaken, but their mere existence demands comment from a philosophical account of valuation.

NOTES

1 The point made here is really a very ordinary one, made by almost all writers on the subject, some more explicitly and in greater detail than others. See, for articles which include some reference to it, Henry D. Aiken, 'A Pluralistic Analysis of the Ethical "Ought," ' *Journal of Philosophy*, 48 (1951), pp. 497–505; Mary Mothersill, 'Moral Knowledge,' *Journal of Philosophy*, 56 (1959), pp. 755–63; D. A. Wells, 'The Psychological Surd in Statements of Good and Evil,' *Journal of Philosophy*, 48 (1951), pp. 682–9. And for an interesting and related point about agreement between competing theories, see S. Cavell and A. Sesonske, 'Logical Empiricism and Pragmatism in Ethics,' *Journal of Philosophy*, 48 (1951), pp. 5–17.

2 R. B. Perry, *General Theory of Value* (New York: Longmans, 1926), p. 115.

IV

FIVE TYPES OF VALUATION

It is easy to see marked differences among acts reasonably described as the valuing of an object. For example, at times, when a man values things, he agonizes over consequences, means–end relations, and what other people think. At other times he does not. Instances of valuing are often quite different in character, and one is led to suspect that any analysis which obscures these differences—for example by trying to show that all acts of valuation are essentially the same—will be impossibly complicated from the outset. For even if it is true that all acts of valuation must have some properties in common, it may not be those common features which are at stake in a given moral judgment. And thus the 'unary' analysis would need to make adjustments by doing the sort of typology done here anyway. In terms of justification-procedure, one must go where the issues are. Differences found among acts of valuation must not be obscured. Moral arguments often center on those differences.

But the attempt to preserve the differences must be a careful one. The dangers of dichotomy are familiar: distinctions, when hardened into a rigid typology, obscure similarities, overlappings, inter-connections. Although valuational acts do indeed differ, it is hard to find cases which do not include feelings or implicit recommendations, just as it is hard to find cases which do not include some cognizance of consequences (or lack of them), and some cognizance of what other people think (or might think) about the valuation being made. The trick is to give an account which does not obscure either the similarities or the differences, and that is a large order.[1] The procedure here will be to point to rough classes of valuations—classes defined by the leading features of various acts of valuation.

The way the class boundaries are marked off is as follows: a thought experiment is undertaken in which one begins with a case agreed to be a valuation and asks of the (imaginary) valuer, 'What is the value of (the object valued)?' When a reply is 'given,' one asks for justification—for a reason to agree—that the object is indeed of that value. *The sorts of reply the valuer can reasonably make to this demand are taken as indications of sorts of valuational acts.* That is, the sorts of justifications which might reasonably be expected from an actual valuer are taken as indications of the leading features of actual acts of valuation. If the justification refers to the utility of the object for some purpose, that is one sort of valuation. If it refers instead to some set of affective responses on the valuer's part, that is another sort. And so on. For convenience, expressions which can reasonably be thought to count as answers to 'What is the value of *x*?' will be called value-expressions.

Perhaps it would be wise to insert a remark here about motor-affective responses: that is, those responses we characterize in terms of feelings or emotions (e.g., pleasure, pain, anxiety, happiness, fear, etc.) and which are correlated with certain physiological changes (e.g., in galvanic skin response, dilation of the pupils, heartrate, etc.). This is an important area for theory of value.[2] Indeed, one is inclined to say that some sort of pro or con motor-affective response (attraction/repulsion; seeking/avoiding; being pleased/being displeased) must be a necessary feature of all valuational acts. That is, if someone were to say that he valued an object but in fact had no pro or con response to it at all, we would probably say either that the claim was mistaken or that it was unintelligible. For the existence of a pro or con response is something we always seem to have in mind when we say a person values an object.[3]

The point to be noted here, however, is that even if the pro/con response is a necessary feature of all valuational acts, it is not always the dominant feature, not always the one emphasized in expressions answering the question 'What is the value of *x*?'—and thus not always relevant to the immediate justification problems raised by value claims. It will, no doubt, always have to be dealt with at some point. But to treat the motor-affective element in valuation as if it were always at the very heart of value claims will be seen, in what follows, to be quite misleading; it is often not the leading issue in the mind of the valuer at all.

It turns out that an application of the procedures just outlined results in the sorting of valuations into at least five rough classes or types. By 'rough' it is indicated that the types are not mutually exclusive, and that their boundaries with one another are often vague. The classes are as follows:

1 SUMMARY VALUATIONS

It frequently happens that when people are asked to justify their answers to the question, 'What is the value of *x*?' they can only (rather lamely, often) cite their estimate of the 'opinions' of others. 'But everyone thinks so!' Or, 'I thought most people agreed . . .' I suggest these replies indicate a very common form of valuation which must be taken seriously if one intends to make headway with the justification question. It is easy to ridicule the sort of act these replies indicate, and to find them pathetic, the poor-of-mind man's substitute for the real thing. (For example, consider the professional's reaction to the common occurrence in a theater offering an experimental piece, when people look around continually to estimate and take cues from the responses of others.) But ridicule does not destroy the reality of the practice, and it might be remarked, as a caution, that the 'voting' procedure indicated here may not always be as ludicrous as it seems. There are tyrannies of the individual conscience as well as of 'mass psychology.' And in democracies, we occasionally think it a sufficient justification for an act that it is approved by a majority of the populace—never mind for what reasons. Even in personal matters, though, the procedure, second hand as it may seem, recommends itself in cases where one's own assessments cannot bring one to a judgment of the thing's value. In perplexity, one reasonably turns to others for guidance. There is, of course, no guarantee that others' assessments will be correct, but turning to them, even accepting them, is recommended (as it would be for one suffering from amnesia) by the admitted limitations of any one man's abilities to assess every matter, and the admitted importance of point of view in getting an adequate, reasoned assessment.

The importance of doing what you think correct, rather than what others say, is properly emphasized in arguments over accountability. ('But everyone says so' is not to be taken as an excuse from responsibility.) But to extend the point to the assertion that one should never act on anything but his own conscience (meaning, not to make up one's mind at all on the basis of others' views) is madness. The appropriate role of authority in moral life is a complicated one—and an issue to which the procedures developed in this book may usefully be applied. Further, the appropriateness of making these summary assessments may depend (among other things) upon one's acquaintance with the object valued. If I am asked about the quality of a movie I have not seen, to give an answer which depends on the valuations made by others—indeed to value the film solely in those terms—is unremarkable; whereas to do so *after* I have seen the film may be quite a different matter.

In any case it happens that people often do cite, as the only evidence for their value expressions, the valuations of others. And I take this to mean that the sort of act they have engaged in to arrive at their value expressions is some kind of rough 'averaging' procedure in which they have somehow decided and at once adopted what they consider to be the dominant view of the matter. ('Dominant' need not mean, of course, 'most frequently or widely held.') In so far as this is all that is going on, one may say the act is a summary valuation.

I have brief hesitation about calling this a genuinely valuational activity; the temptation is to say that it is rather an operation on data which are valuations. But the hesitation is only a brief one. The procedure *is* an operation on valuations, but to deny that it is also a valuation in itself seems misleading, since it is undertaken in circumstances in which it is sought as an answer to the question, 'What does the summarizer say of the value of x?' rather than an answer to the question, 'What do people other than the summarizer say of the value of x?' The latter question is the sociologist's. When the summary is undertaken in the context of the former question, however, we need not hesitate to call it a valuational act. If the summarizer's purpose is to find out what he thinks by finding out what others think, then it is hard to see how the title could reasonably be refused.

The results of such assessments can be attended with all the motor-affective responses involved in the more standard examples of valuation, and people often act from the basis of these summaries with the same confidence and in the same ways that they act from, for example, their judgments of utility. They may, in fact, be much more confident in their assertion of a thing's value if they have come to it by finding out what 'the group' thinks than if they have come to it on their own.

As for the meaning of sentences expressing summary valuations, apart from the summary, one would expect them to carry some hint (however small) of (1) approval or disapproval; (2) readiness to act for or against, toward or away from; (3) attraction or aversion; (4) impulse or expulse; (5) acceptance or rejection. . . The list of possible pro/con responses is a long one. But it is clear that to relate the meanings of 'good,' 'bad,' and variants thereof exclusively or even primarily to these motor-affective aspects is misleading for the purposes of an inquiry into justification procedure. For the fact is that in this type of activity a good deal more attention is given to the summary than to whatever motor-affective response attends it. While the presence of some affective response may be necessary if we are to call the 'summarizing' properly a valuational act, it is still the case that the dominant feature of certain of these experiences is just the summarizing. And I suggest that, when it is argued that what is 'essentially' valuational here is the affective or calculative response rather than the summary, one tends to lose sight of the experientially dominant features of the activity, to misrepresent the emphasis in the meanings of statements associated with it, and to end in confusion over what is to be justified (finding the evidence cited by the valuer to be irrelevant, even ludicrous).

2 AFFECTIVE VALUATIONS

An even more frequent reply to the demand for a justification of value expressions than the one noted above is the one which cites feelings, a 'sense' of a thing's value, an 'intuition' of it, or simply some motor-affective response of the pro/con type previously mentioned. The distinguishing feature here is that when one asks, 'What justification can you give for saying, "x is of y-worth"?', the reply simply cites some personal awareness, or affective response. (If the reply claims that the object valued possesses some value-*making* property, in virtue of its ability to please, and so on, the valuation involved is best construed as calculative [see below] rather than affective.)

With some hesitation the activity indicated by the sorts of answers just outlined will be called affective valuation—not because intuitionist and moral sense theorists' claims to have observed some objective property constituting value can be disproved, but because (as Moore himself noted) no conceivable evidence could be advanced in favor of them, and as yet the only commonly-agreed feature in all of these 'no evidence' views is the presence of the affective response.[4]

To explain: the weakness of intuitionist and moral sense theories (for present purposes) consists in the fact that no criterion has yet been offered for distinguishing reality from illusion in the case of the awareness of objective value. In the face of widespread and otherwise irreconcilable disagreements, this is most dismaying. A sighted man cannot 'prove' to a blind man that he (the sighted man) can see; any evidence offered (better mobility, etc.) can always be given another explanation by the blind man. Similarly for sighted people trying to prove that they have sight to other sighted people, and so on. The evidence will always be ambiguous with regard to its interpretation. But it is crucial to note that these half-way proof procedures cause no trouble in practice; there is a lack of significant disagreement in the human community as to who is blind and who is not. Cases are ordinarily so clear cut as to rule out whole batches of alternative explanations in favor of the probability of the ones agreed upon. But no such luck is to be found with the question of who is in error in sensing or intuiting a value—of who is sighted and who is blind. And in the absence of either agreement or a criterion, the justification question is hopeless. I am personally convinced (but of course cannot demonstrate) that no such criterion, and no such agreement, will ever be reached. So the focus of my remarks on this form of valuation will be directed to its affective aspects only. The procedures developed in this book, however, could accommodate the discovery of the long-awaited criterion, or the long-awaited agreements.

Until then, however, this much can be said for the activities indicated by this group of answers to the demand for justification: there is a kind of activity whose distinguishing feature is simply the awareness of a pro or con affective response to a 'thing' ('thing' again in the widest sense). Perhaps the most compact description of this sort of valuation, although one heavy with jargon, is given by J. E. Reid: 'I shall mean . . . immediate awareness, qualified by positive or negative hedonic tone, of a massive, unanalyzed *attitude response*, liberating a variable amount of nervous energy and involving both cognitive and motor factors.'[5]

> Example: 'I don't know, I just feel uneasy about it. I have the feeling it is evil.'
> Example: 'It's good because it is symmetric and I like symmetry.'

The examples illustrate two forms of the activity (i.e., they are meant to amount to expressions of the outcome of two forms of the activity). The first expresses a response to the thing as a whole; the second expresses a response to some property of the thing. Neither, it should be noted, makes reference to the function of the object (or property) except in so far as calling forth some affective response from the subject is held to be its function. The affective response may, of course, involve some inclination to act as a consequence of the response. (It should be noted, also, that such valuations need not always be a response to the object valued—at least not in the everyday sense of 'response'—but may also be a sort of gratuitous output of affection directed toward the object.)

This is a very significant form of valuational activity, and under the conditions imposed above on the present discussion, it must be regarded as a form of 'immediate,' 'non-cognitive' valuation of the 'intrinsic' value of an object, and a valuation which has all of the 'indisputable,' 'culture-relative' properties mentioned in Chapter III. To explain: (a) The affective valuation is a reference to 'intrinsic value' in the sense that no reference to the object as a means is made (beyond the occasional but not necessary reference to its 'capacity' to produce the affective response). (b) It is non-cognitive in the sense that—and in so far as—it is merely an affective response. It might be argued of course that even an affective response can involve cognition in at least three senses: in so far as one 'recognizes' it as 'an' experience; in so far as one knows how to express it so that others understand; and in so far as one knows that, and how, the response was 'caused.' But in that case, I know of no experience which is non-cognitive. (c) The affective valuation is immediate in the sense that, though 'becoming aware' of the affective response might take considerable effort (concentration), it is not the product of a calculation of consequences. I do not mean that an affective response may not result from such calculation, merely that the activity presently under discussion is not calculational. Activity which is calculational will be discussed below. (d) The affective valuation is indisputable in the sense that one's report of 'evidence' for the value expression (supposing he knows the conventional words to use) cannot possibly be mistaken. In the symmetry example above, it is the *response* to symmetry which is used as 'evidence' for the valuation, not the 'fact' of symmetry itself. And I cannot legitimately be told I do not know what my response is. I may, of course, be told that my response is morally wrong, but that is another matter. So in this sense there is indeed no disputing taste—no point in saying a man's valuation is incorrect. The value is 'in' the response of the valuer.

What the words mean, when people express their affective valuations, is clearly a very broad question. Within the limits imposed by the general usage of 'good' to indicate the 'pro' response, and 'bad' to indicate the opposite, it seems clear that there can be no *a priori* reason for supposing that the value expressions always must indicate some one sort of affective response, that they must always commend or command, that they must always indicate approval in the usual sense. These items are not necessary elements of a 'pro' response. One can find a thing compelling and express this as a valuation of the thing without wanting to say, one way or the other, whether he approved of it or not, and without wanting to commit himself to commending it to others. Unless rather technical definitions (widening ordinary usage) are given to 'approval' and 'commendation,' these elements cannot be seen as necessarily involved with valuation. They can more reasonably be seen to be involved with *e*valuation, however, as will be noted below.

In any case, to deny this flexibility to the meaning of value expressions is to miss the point, which is that the discoverably dominant feature here is simply the awareness of the pro or con affective response. The responses which legitimately can be verbalized with 'good' and 'bad' (i.e., which fall within the general limits mentioned above for 'good' and 'bad') are as diverse as sadism and masochism; the mechanisms which produce those responses can be as unrelated to the properties of the objects valued as those of the paranoid or as dependent on the properties of the objects valued as those of the person who likes music only in the key of C. It is, after all, possible and sometimes important to indicate, at least in expressions *about* one's affective valuations, whether or not his response is thoroughly dependent on a particular property in objects, or is governed by a strong tendency toward masochism, or is totally free of any 'objectification of emotion,' or totally involved with insistence that others see things as he does. One may say, of course, that in so far as affective value expressions are expressions *of* the awareness (rather than statements *about* that awareness) they are not, strictly, assertions (statements, judgments). This distinction between expressions *of* and expressions *about* affective valuation will be discussed in more detail when preparations are being made to incorporate value-expressions into evaluative procedures.

3 CALCULATIVE VALUATIONS

I suspect that the most frequent reply to the demand for the justification of one's value expressions is to cite some means-end relation that the object has: 'Why is it good? Because it can do so many things: sharpen knives, open cans. . .' And the activity 'behind' such replies is the familiar one of estimating what an object is good for, its instrumental value. Perhaps the most frequently practiced valuational activity, then, is the estimation of what an object is 'good for'—that is, the estimate of its effectiveness (elegance, simplicity, 'fecundity') as a means. I propose to call this activity calculative valuation, subject to some cautionary remarks. (1) It may occur in *re*spect to a purpose, or in *pro*spect of one; it may also occur in *retro*spect. (2) The word 'calculative' may be a bit misleading if one supposes it to mean a detailed and validly reasoned procedure. Such procedures often occur in—or sometimes constitute—the calculations, but they are by no means always present. Whether this is lamentable or not is left for another time. 'Calculative' is thus used very broadly to indicate any way of coming to a decision on the matter of one thing's 'effective relation' to another. A decision on this issue can be made by a process akin to affective valuation as described above. One gets or invites an affective response to the thing considered as a means. In this case, the justification for using the title 'calculative valuation' may be slim, but not altogether absent.

A crucial way in which this procedure is distinct from affective valuations, however, is that the calculational result is often falsifiable. One can, of course, construct purposes with respect to which a thing's effectiveness cannot, at least presently, be tested unambiguously. 'Flagellation is good for getting into contact with the devil' is one such valuation. But very many calculations are straightforwardly falsifiable, and these are cases in which we can understand the claim that the nature of the thing valued, combined with the nature of the purpose in terms of which the thing is valued, 'determines' which valuations are mistaken and which are not. The more thoroughly one considers these items in calculating a thing's value, the more thoroughly reasoned is his calculation. Its falsifiability, however, is not determined by whether it was reasoned or not.

(3) The motor-affective aspects of calculative valuation can be explained, at least in part, as the consequences of whatever 'striving' is involved in the process—whether this striving is hypothetical or not. That is, there is a kind of impetus given by a purpose which, when helped along or hindered, is reflected in consequences for our affections. Precisely how this comes about is a question for empirical psychology; precisely what consequences will occur in a specific individual for specific situations is a question to be answered in that person's psychoanalysis (lowercase 'p'). But it is clear that mild affective responses occur even when one calculates prospectively (or retrospectively) and finds that a thing helps or hinders some hypothetical (or past) purpose.

(4) What, then, do sentences of the form '*x* is of *y*-worth' mean when they express calculative valuations? As might be expected at this point, the answer will be that the sentences can mean a great many things. After all, what has been called the 'efficiency-relation' can be understood in a variety of ways: as a thing's utility for one and only one purpose (never mind finesse, never mind other consequences); as a thing's potency, not only for the purpose at hand, but in terms of its fecundity with respect to

generating further opportunities for action or satisfaction; as a thing's efficiency and elegance; as a thing's necessity and pleasure-giving potential. . . The possibilities (after having counted four) seem endless. Further, the variety of ways in which the calculations can be done has been indicated; one may want to make reference to the method in reporting the result. Finally, one may want to convey something of the affective response consequent to the calculation. A man says he 'may want to' convey this, but it may be unavoidable if, and in so far as, the responses themselves are unavoidable. Here again the question of the omnipresence of the pro/con affective response is involved. A calculation without such a response might not be admitted by many to be a valuation. In any case, when one lists the possible combinations of the three items ordinarily in the meaning of expressions of calculative valuations (the sense given to 'efficiency relation'; the manner in which the calculation was done; the particular affective response consequent to the calculations), he can see the difficulty of presenting any further analysis of what the expressions must always mean.

4 GOOD-OF-ITS-KIND VALUATIONS

It frequently happens that, in answering the demand for a justification of their value-expressions, people cite some standard for the type of thing being valued: 'But don't you see, if it were a really good hammer it would be. . .' where the ellipsis is filled in, not with reference to what it is or is not good for, or what it does or does not cause one to 'feel,' or what people in general believe about it, but by reference to what the hammer would be if it came up to or exceeded the standard of what it is to be a hammer.

The way in which one fills the ellipsis here is crucial, for utterances which, on their face, appear to be good-of-its-kind valuations, may turn out on closer inspection to be something else. For example, they may be better analyzed as calculative valuations—e.g., a good hammer is one which gets my hammering done. Or they may be better analyzed as affective valuations—e.g., 'That's a wonderful sunset.'

But sometimes the crux of the matter is *just* the object's relation to some standard—never mind what 'reasons' (calculative, affective or whatever) we have for adopting those standards. Consider 'That's a wonderful sunset, as English sunsets go.'

When this is the crux of the matter at stake in a justification procedure, it seems best to deal with it directly—by marking it off as a special sort of valuation. The valuation of the 'kind' or standard can be dealt with in the consideration of evaluative procedures.[6]

The occurrence of this sort of valuational act depends on the sorting of objects into types, for which either adequate type-descriptions can be given, for which some other form of decision procedure is available for doing the sorting. Then the valuation of the object as good (or bad) of its type can be given either (1) in terms of the ways in which the object exceeds or falls short of the type-description; or (2) in terms of the significance of any objections to the 'success' of the object as one of its type.

In the first case, an object's meeting the type-description is just that: the zero point on the value scale. To be valued as a bad hammer, some objection must be entered against it: 'Well, it's a hammer all right, but look here, the handle is split. It will break any minute.' To be valued as a good hammer, some special excellence must be claimed for it. But deficiency and excellence here are usually extrapolations of elements of the type-description: hammers have handles, both for leverage and the degree of ease they afford. Thus, although *x* may have a handle, it may be objectionable *qua* hammer-handle.

The second case—much more frequent because of the difficulty of nailing down precise and adequate type descriptions—develops in the same way from an initial agreement about what type of thing the object is (never mind why it is what it is—that is, what set of necessary and sufficient-type-conditions it meets), and proceeds to a valuation of the object in terms of its flaws and virtues.

Before moving on, however, it might be wise to deal with the objection—raised, for example, by Paul W. Taylor in his book *Normative Discourse*[7]—that good-of-its-kind value expressions cannot be *value* judgments at all because they do not always express a pro or con attitude compatible with the positive or negative nature of the expression. For example, one may say something is a good torture instrument, but precisely *because* of its being good-of-its-kind, one is horrified at it. It is not considered 'good on the whole.' Two things need to be pointed out: first,

Taylor has conflated the valuative and evaluative acts (it is notable that in all the crucial places he speaks only of *e*valuation). A valuation results (when expressed) in a judgment on whether an object is good, bad, or neutral. An evaluation is an attempt to 'balance' the various goods and bads to determine what the object's value is 'on the whole.' And so because in his example the good-of-its-kind expression does not express what is 'good on the whole,' he is prepared to say it is not being used as a *value judgment*.[8] Denying it this status seems arbitrary and misleading when one is prepared to describe it as expressing 'good with respect to some standard' (as Professor Taylor does) and further note the distinction between valuing and evaluating (as Professor Taylor does not).

As to the second point Taylor's argument raises, it is not at all clear that (again where the valuative and evaluative levels are distinguished) one may not be horrified at torture instruments (the affect associated with the evaluation) but have, as we say, 'a horrible fascination' or a 'grudging or disturbed admiration' for the rack as good of its kind. In short, the conflation of the valuative and evaluative acts results (not surprisingly) in an oversimplification of human value experience—a very standard part of which is the occasional 'horrible fascination with' or 'grudging admiration of' an object we regard (on the whole) as vicious in the worst imaginable sense. Such complications of the value experience are by no means necessary, as one may well not make good-of-its-kind valuations of the rack or whatever. Indeed, casuists discourage us from doing just that. And it may be a justifiable moral rule to hold that one should refrain from diluting the force of overwhelmingly negative *e*valuations by not reflecting on any excellences the objects involved may have. But that is a matter quite beside the point at issue here.[9]

5 FITTINGNESS VALUATIONS

There is, finally, another sort of activity referred to as a basis for concluding that an object is good or bad which can only be described as the process of determining the object's fittingness. This activity is characterized by attempts to determine whether or not the object is 'compatible' with its surroundings, 'consistent' with some standard, or to use some clichés, a 'good match,' 'in tune with the times,' 'out of step.'

One hesitates to class the activity as a version of affective valuation. Clearly what counts as a 'fit' is often decided, in part, by aesthetic standards which make use of feelings, but consistency and compatibility also involve matters open to falsification. One hesitates, too, to class the assessment of 'fittingness' as a version of calculative valuation—even though the 'fittingness' of an object often bears on whether or not it 'works' (i.e., is good *for* anything). Often, fittingness has little if any relation to means-ends considerations. And one hesitates to class the assessment of fittingness as a version of good-of-its-kind valuation, because often, to say that *x* is suitable, or 'fitting,' is not really to claim any special (i.e., species) excellence for it at all.

In short, the assessment of the fittingness of an object seems separable enough from the other forms of valuation to justify treating it as a separate type.

6 TWO OBJECTIONS

Each sort of valuation leaves questions unanswered. If one's claim that *x* is good is a summary valuation, the critic will want to know what sort of valuations were contained in the summary—and what made *them* correct. If one's claim is an affective valuation, the critic may say, 'So what? How does that license any conclusion about what it is right to do?' For calculative valuations, one wants to know the value of the thing *x* is 'good for.' For good-of-its-kind valuations, the issue is the good of the kind. And for fittingness valuations, the concern is for the value of fitting (in this or any case). These 'second order' questions will be taken up in the next chapter.

But at the moment, there are two sorts of objections to the typology outlined here which must be dealt with. One has to do with clarity. It may be objected that crucial terms (such as 'awareness' of an affective response) are vague, that the principle by which these types of valuational act were picked out remains something of a mystery (are there more 'types'? how can one be sure?), and that the reasons for sweeping aside more traditional accounts of valuation are still somewhat obscure. That is one type of objection. The second and perhaps more serious, which I will call the Socratic objection, has to do with the grounds for insisting on a genuine plurality of types of valuational act. After all, if we are to treat all these acts as in some sense 'the same sort of thing' (i.e., valuations), does that not mean that however different they may appear (be) 'phenomenologically,' they are all, as valuations, essentially the same—that they have some characteristic or set of characteristics in common without which they could not be valuational acts? Is not this the 'essence' of valuation, and of necessity one thing rather than many?

THE OBJECTION CONCERNING LACK OF CLARITY

One need not attach too much importance to the particular descriptions, and especially the titles, so far given. The purpose has been *to point*, to give an ostensive rather than discursive indication of the regions of activity under discussion. No argumentative subtleties will depend on having chosen terms like 'awareness of an affective response' or 'calculative valuation' rather than some others. What I assume (because it appears to be correct though I have no idea how to 'prove' it) is that the activities to which my words point are themselves sufficiently distinct from others so as to render complete discursive precision (whatever that might mean in this context) needless—at least initially. As long as the words used are not misleading, no harm will result from beginning with this simple technique. The reader will note that some care has been taken to delineate what the words do not mean. This caution, together with the fact that the activities to which the words refer are familiar, means that some vagueness in expression will not interfere with the reference made.

The way in which the list of valuational acts was arrived at is a straightforward, commonly-used procedure. One asks what sort of activities people are willing to call valuational and each of these activities is taken to be a valuational act until there is evidence to the contrary. What would count as evidence to the contrary? Well, of course, finding out 'what people are willing to call' a valuational act is one of those tasks which in principle requires a great deal of empirical data collection, but which philosophers have traditionally barged through on the basis of their personal understanding of the language and the help of lexicographers and tradition. I have, I fear, done neither better nor worse than that. An account of how this procedure works would go something like this: There are two basic problems: (a) How do I know that an activity is really an example of valuing? (b) How do I know that an example of valuing is 'really' what it seems to be?

(a) Suppose someone points to an activity (A) and says, 'That is an example of a person valuing something.' I, however, am not sure if he is right. How do we settle the question? I have at least two alternatives:

(i) I can propose an alternative description of (A): 'No, no, he's playing tennis, you fool.' And then we can argue about the proper use of words. Or

(ii) I can try to spell out why I am hesitant about describing (A) as 'valuing something': 'But I thought that, in order to value something, you had to be paying some attention to it at least—and he's asleep.' Then we can argue about the proper use of words. In short, disputes over whether x is or is not 'really' a valuative act are always properly posed with the question: is x properly called a valuational act? The most obvious attempt to settle this question is, of course, in terms of the appropriate language conventions. How do people use the term? When would they be willing to call x a valuational act?

(b) But after the question of an activity's 'properly being called a valuative act' is settled, in terms of conventions regarding when people would call it that, there still might be disagreement about why it is properly called that or disagreement as to its precise description. On the matter of the precise description of the act, it has already been noted that the procedure was to assume that the nature of the activity is indicated by what sort of replies people make to demands for a justification of their value expressions. That is, that the sort of evidence (or non-evidence) one cites is a reliable guide to deciding what sort of activity was involved in the valuation. At least two objections are possible here: one that this procedure might not be discriminating enough, papering over remarkably distinct acts with the same label; the other that it is too discriminating—making more of differences than is necessary. Neither objection can be answered within this book since only the success or failure of the enterprise as a whole can decide these matters. It is hoped, of course, that both objections will be met by the book. One further comment: it might be noted that very often value-expressions occur in a way such that the demand for justification startles the speaker into 'thinking one up.' In such cases, the evidence he cites does not indicate the sort of activity he was in fact engaged in, but rather the sort he might have been engaged in, or which he thinks he ought to have been engaged in. For the purposes of finding the varieties of valuational act, this feature has only a procedural significance for the selection and interpretation of cases.

If, once the act is agreed to be a valuation, and its description is agreed upon, one still insists on knowing why each act is called a valuation, this insistence can only be interpreted as a request for further discussion of the linguistic conventions (which is an unlikely interpretation), or an example of what will be discussed below as the Socratic objection. Further discussion of the linguistic conventions can really only repeat what has been said already—that the way one goes about finding out what people are doing when they are valuing things is to look at likely cases—cases in which people use value words, for example, and where other descriptions of their use of these words do not seem to fit (i.e., they are not playing a parlor game, or sending a coded message or being silly or sarcastic, etc.), and one seems forced to say that they must be valuing the thing.

Now as to disagreements with or disproof of this sort of language analysis, the counter-evidence will again be 'linguistic' in nature. Such evidence (at least *prima facie*) would be constituted by genuine unwillingness to call one or more of my types valuational in character (or willingness to call something a valuational act that I cannot in good conscience so describe). What happens if general agreement cannot be reached on these matters? If no amount of explication, no attempts to state necessary and sufficient conditions for an act's being a valuing are capable of resolving the disagreement? If, in short, there are irreducible differences of opinion between equally proficient speakers of the language? Or evidence to suggest that in some other language (and/or culture) quite a different list of types of valuational act would be given? In that case, the whole of the general theory of value presented here (and all consequences proceeding from it) becomes conditional in logical form as well as provisional as a matter of intellectual honesty. The form then becomes: If one accepts this list of valuational acts rather than lists *x*, *y*, and *z*, then consequences (A) follow, rather than consequences (B), (C), and (D) respectively. In other words, irreducible disagreements at this point complicate the task, but do not stop it, and do not in principle entail irreducible incompatibilities in the sets of consequences deduced from the various lists of valuational acts. So agreement on the typology, while certainly desirable, is not strictly necessary to getting an effective set of justification procedures.

It might be appropriate here to cite, by way of illustration, one possible candidate for inclusion in the typology which, after some consideration, was not included. It might be thought that the assessment of *economic value* ought to be roped off as a separate form of valuation, for when asked for the value of an object, a man may well say something about its dollar value. If this turns out to be admissible, I would not be in any way distressed, but I have not treated 'dollar value' statements as a special type here because it seems to me that assessments of economic worth are usually classifiable either as summary valuations (as when I, who know very little of mechanics, ask for the Blue Book price on my car), or as calculative or fittingness valuations (as when I am pushed for some explanation of why my car has a given dollar value). The point is that in assessing economic worth, I often only am concerned to find out 'what the market price is' (summary valuation). But that when I am pushed for an explanation of the market price, or exchange value, I usually give an answer in terms of usefulness (calculative valuation)—or again, in terms of convention (fittingness valuation).

In any case, it should be obvious from this simplified review that the procedure in marking out the types of valuational activity has been in the main a standard one (albeit that standard procedure has certain outstanding difficulties). There is one facet of it, however, which is a bit of a departure from the standard, and that is that once the list of activities 'people are willing to call valuational' was compiled, rather than search for the common features of all such activities, the arguments simply pointed to the 'whole' of each activity as valuational. To explain why this was done is, in effect, to speak to the second objection mentioned above.

THE SOCRATIC OBJECTION

Two theses related to the matter at hand are now widely acknowledged: first that it is perfectly possible (given the haphazard way natural languages grow) to find several activities having the same generic name but having no significant 'distinguishing' properties in common; second, that it is very often (perhaps almost always for significant genera) a hopeless task to try to specify a set of necessary and sufficient defining conditions, or an 'essential' or 'real' definition which covers every case. 'Family resemblances' are usually the most that can be hoped for.

If several things called valuation have no significant property in common, it is clear that we would mislead ourselves by lumping them together for the purposes of moral philosophy. We would, in that unhappy case, have to (1) take them all but treat them very differently (there being, then, no general theory of value); or (2) take one (or more) which do have significant features in common in order to have a general theory of value. In either event, the pluralist's case seems to be lost: in the first because he can show no plurally *general* theory of value; in the second because he can show no *plurally* general theory of value. In the second case, it appears that the similarities or defining conditions are the crucial features, and that the general theory of value is again a theory of one (or one set of) thing(s). If one can specify family resemblances it can be argued that these resemblances indicate a kind of unity —and that it is the resemblances, rather than the 'whole types,' which ought to be emphasized.

These arguments carry weight; indeed, they are probably behind the tacit refusal of most writers to consider a pluralist theory of value. But these are not arguments against the kind of procedural pluralism adopted here. The arguments of this book simply call attention to the fact that the uses one can legitimately make of the values one arrives at depend largely on the nature of that valuational process. And there is no guarantee that the uses we will find it important to make of values in moral philosophy will all be related to (legitimized by) the 'essential' properties or 'family resemblances.' In other words, while finding out what all valuational acts have in common may be interesting, while looking for the family resemblances and/or the 'essence' of all valuational acts may be good metaphysics, it may lead us astray in moral philosophy. For we may find that these common features, these family resemblances—even the essence, if there is one—may not always be the operative features of valuational acts from the standpoint of the justification question. Thus our concentration on the essence may persuade us to view very ordinary operations with valuation as peculiar, unwarranted, or sheer nonsense merely because they are not, as such, legitimized or possible from the point of view of the 'essence' alone.

For example, if we take a certain motor-affective response of attitude orientation to be at the heart of all valuational acts (as I suspect it is) and then go on to say that that 'heart' is what it 'means' to value something, we quite naturally take a dim (or at any rate extremely complicated) view of talk of the warrantability of value judgments. How can 'taking an attitude' be true or false? But if, on the other hand, we pay attention to the whole of the act called valuational, we may more easily see in what senses evidence bears or fails to bear on what is being done.

This is just an example. The point is that one suspects moral philosophers have been seriously misled by the 'Socratic quest' in describing valuation.[10] The intent here is to keep the whole of each of the several valuational acts in view, in the hope that this tactic will be seen to simplify and clarify many of the old, and genuine, quandaries in value theory.

NOTES

1 See, for discussions of the problems of a typology of values: Edward Schuh, 'Syntax of Inherent Value,' *Journal of Philosophy*, 53 (1955), pp. 57–63; Alan Gewirth, 'Meaning and Criteria in Ethics,' *Philosophy*, 38 (1963), pp. 329–45; P. B. Rice, 'Toward a Syntax of Valuation,' *Journal of Philosophy*, 41 (1944); E. Gilman, 'The Distinctive Purpose of Moral Judgments,' *Mind*, 62 (1952), pp. 307–16; W. A. R. Leys, 'Types of Moral Values and Moral Inconsistency,' *Journal of Philosophy*, 35 (1938), pp. 66–73; P. B. Rice, 'Types of Value Judgments,' *Journal of Philosophy*, 40 (1943), pp. 533–43; R. M. Millard, 'Types of Value and Value Terms,' *Journal of Philosophy*, 46 (1949), pp. 129–33.

2 R. B. Perry, *General Theory of Value* (New York: Longmans, 1926), and Stephen Pepper, *The Sources of Value* (Berkeley: University of California Press, 1958), both have elaborate analyses of motor-affective responses as they are involved in valuation.

3 When I speak of 'what we are willing to call a valuational act,' I do not mean to indicate the results of a survey of 'man on the street' opinions. Obviously, 'valuational act' is a rather special locution. Many people would merely be puzzled by it. But in circles where it (or close variants) are familiar—that is, in precisely those circles from which a lexicographer would draw his definitions of 'valuation'—the phrase 'what people are willing to call a valuational act' makes perfectly good sense.

4 'But surely,' one hears, 'it will not do to lump values *for* objects (subjective, emotive, etc.) with values *of* objects. Surely if one can exhibit the values of

objects, that is the crux of the matter for moral philosophy.' Not so. Besides the rather unhelpful distinction between values of and values for, the plain fact is that moral judgments can be made about either type and can continue to be made about values for objects even if an ironclad way were found of identifying the (objective) values of objects. So the procedural question must consider all valuations. This matter is driven home, in similar language, by Nicholas Rescher in Chapter V of his book *Introduction to Value Theory* (Englewood Cliffs: Prentice-Hall, 1969).

5 J. E. Reid, 'A Definition of Value,' *Journal of Philosophy*, 28 (1931), pp. 673–89 (in particular pp. 673–4).

6 For an interesting attempt to construe the meaning of 'good' along lines very similar to these, see R. S. Hartman, 'A Logical Definition of Value,' *Journal of Philosophy*, 48 (1951), pp. 413–20. More recently, Dorothy Mitchell, 'The Truth or Falsity of Value Judgments,' *Mind*, 81 (1972), pp. 67–74, uses the good-of-its-kind notion prominently. I am indebted to Charner M. Perry and Alan Gewirth for pointing out the difficulties in treating good-of-its-kind valuations as a separate type.

7 Paul W. Taylor, *Normative Discourse* (Englewood Cliffs: Prentice-Hall, 1961), pp. 54 ff.

8 *Ibid.*, p. 54.

9 Just as an addendum, it might be noted that some agent-morality theories seem thoroughly involved with good-of-its-kind valuations—in the sense that standards of excellence for humans *qua* humans are used as criteria in terms of which priorities are assigned to various virtues. But proposals to be developed later in this book will show, I think, that there is a much stronger way of organizing agent-morality considerations.

10 For some polemical remarks directed against Moore's 'essentialist' stance, see Stephen C. Pepper. *The Sources of Value* (Berkeley: University of California Press, 1958), Chapter 1, section 3. Pepper's remarks in Chapter 1, section 2, are very much the sort of cautions I have been advancing here, though he is more concerned with description than with justification procedure.

V

MORE ON VALUATION

The arguments so far have only served to introduce the five sorts of valuation. Several serious questions about them must be dealt with before we move on to see how they fit into justification procedures. One wants to know more, first, about the process involved in each way of valuing things. One wants to know whether some of them might not be 'reducible' to others. And one wants to examine certain second-order questions that arise for some of the types.

1 COMMENTS ON THE PROCESSES INVOLVED IN VALUING

On the matter of the way in which one arrives at valuations of each type, little needs to be said about summary and calculative valuations. Procedures for sampling attitudes and opinions, and procedures for finding out about one thing's means-end relation to another are standard and in no way special to value theory. The pro/con response takes care of itself. It either tags along with the procedure or it does not. The 'adoption' of the valuation obtained in either way is not properly thought of as a separate act, requiring some explanation, but rather as simply being the result of the endeavor: that is, the whole point in collecting opinions or finding out about the efficiency relation is to 'get' the valuation. When those things are accomplished, that constitutes the valuation.

'Overall approval' of the thing, or the inclination to recommend it to others, may have to await the results of evaluation, however.

As for affective valuation, the procedures here are either unobservable (and thus indescribable), as in the case of intuitionist and moral sense claims, or else they are in the domain of psychology. In any case the nature of the causes of the affection are not really at stake when the 'objectivist' position is ruled out. For the 'evidence' for the value expression is the existence of the affective response, not its origin. And, again, ruling out the objectivist view in this case is only that—in this case. It does not mean that the objectivity of value cannot be discussed with respect to other sorts of valuation. Further, this ruling out is, again, only done because of the unavailability of an adequate account of the position(s) for present purposes. Nothing in the arguments on justification procedure will hang on the omission in a significant way.

The procedure for securing a good-of-its-kind valuation is difficult to describe in a detailed yet general way. Disagreements over valuations of this sort can generate rather complex arguments, each full of difficulties special to the particular case, or at least to the particular type of object under discussion. Other than that, however, the procedures are not remarkably unique. Agreements on type-descriptions and subsidiary definitional items are mostly matters of convention, and once these are decided, two questions remain.

The first is to determine in what respects the object exceeds the requirements of its type, or in what respects it is deficient. Answers to these questions can be given by standard observational methods. Observation need not be treated quite as restrictively as an operationalist might insist upon: e.g., in assessing whether an aesthetic experience is good-of-its-kind, one can be said to observe and compare experiences. Or, if not, we shall have to say that some things which can be 'typed' are not available for good-of-its-kind valuations. Either way, the procedural point here is not affected.

The second matter concerns 'degrees' of goodness—the relative weights to be given to the various deficiencies and excellences found. No strict calculus may be possible here, of course, since one may not be dealing with objects which meet all the formal conditions for arithmetic averaging procedures. But a good deal of help (i.e., criteria for fixing the relative importance of deficiencies and excellences) is given by the features of the type-description itself. For example, if a hammer is defined as 'a carpenter's tool which. . .' then it is clear that considerations of utility strike more directly and with more force in this case than do aesthetic considerations. It is not that aesthetic considerations are forever irrelevant to the value of a hammer as a hammer. It is only that they can claim no title to relevance from the type description alone, whereas utility can claim relevance from the description alone. One need not even say that considerations with a title to relevance have a stronger claim, or must be considered more important (for establishing that might be a bit difficult). One only needs to let the untitled considerations fend for themselves; if a claim for their relevance can be established, then they should be treated as all other relevant considerations. That means that unless there is a specification of the relative degree of importance in the title to relevance (derived either from the type description or some other source) they are all taken to be of equal importance. And in that case the quantity of excellences determines the matter. One need not insist, in striking such a balance, that the items being balanced exhibit all the properties of the real number system—or even all the properties of additivity. The matter will be discussed further in Chapter VI.

A consequence of this procedure is, of course, that very many cases cannot be decided. One cannot strike a balance for some reason, or get an accurate line on what counts as a deficiency and what an excellence, and so on. But this simply means that no valuations of this sort can then be made. And even if it were to turn out that none of this sort could ever be satisfactorily made, it would only mean less work for the justification procedures to be outlined from this point on. It would not constitute an objection to those procedures or indicate that they were in some way deficient.

As for the steps in determining fittingness, one can only remark that the standards to be used are given as much by the situation (i.e., the needs and wishes of the valuers involved) as by any context-free features of the objects valued or of the notions of 'compatibility,' 'consistency,' etc. For when diversity is what is wanted, what 'fits' into what one already has is not something that is consistent in the sense of *similar* to anything one already has. Examples could easily be multiplied. The point is that the process of determining fittingness usually proceeds simply by making a set of (more or less temporary) conditions explicit, and then comparing the object to those conditions. And arguments over the fittingness of various objects often take the form of negotiations over choice of those conditions: are we to think of this room as having a 'formal' atmosphere (in which case the magazine rack is out of place)? Or . . . Arguments over fittingness valuations, then, often concern matters of the second order: the appropriateness of the conditions established in each case for defining what is fitting.

As a further comment on the processes described here as valuation—of whatever type—it should be noted that in so far as the results of these valuations enter into disputes about the relative weight of one value as against another, *so far they are all on an equal footing*. That is, as yet the question of the rank ordering of various values has not even been broached. So far, all we may say is that no value for an object (pro or con) *prima facie* outweighs any other value for that object. In order to decide, for example, whether my affect can be more decisive in deciding what to do than your calculative valuation, one must go through the process I describe in the following chapters as *e*valuation. Of course in practice we do not always or perhaps even often separate the two stages. We may conclude in the process of valuing an object in some way, calculatively, for example, that we will not be swayed by any contrary valuations of other sorts or from other sources. But for the purposes of analysis, and thence for a rational justification procedure, it is convenient to separate the process into two 'stages': valuation, in which we simply arrive at pro or con judgments by the processes described as types of valuational act; and evaluation, in which we try to sort out these values into some sort of rank order, or priority-listing so as to be able to weigh conflicting values one against another in deciding what is good or bad 'on the whole.'

2 THE INDEPENDENCE OF THE FIVE TYPES OF VALUATION

Another question which the typology of Chapter IV raises is

whether all five sorts of valuation are independent of one another —that is, whether one might find that there were really two (or perhaps, one, or three) fundamental types, from which the others derive. One would welcome any such reduction. But a formal proof of the independence of the five sorts of valuation seems impossible in principle. (At least, this writer has found no way to do it.) The most one can do is to show that people cite five sorts of 'evidence' for their value-expressions. But then the issue merely becomes, 'Are the sorts of evidence really distinct?' They appear to be, and they will be so regarded here. But again, the adequacy of the procedures to follow—and the usefulness of the approach to moral philosophy through the procedure question—does not depend on the outcome of this issue.

3 SECOND ORDER QUESTIONS

One must note carefully the necessity exhibited by each sort of valuation for a ground, defined as a non-arbitrary stop to reason-giving. Four of the forms of valuation have an 'unfinished' character, and the fifth has an apparently irreducible arbitrariness. In the case of summary valuations one wants to know about the appropriateness (correctness) of the valuations that have been summarized. In the case of calculative valuation, one wants to know about the value of the thing x is good for. For good-of-its-kind valuations, one wants to know the value of the kind. And for fittingness valuations, one wants to know the appropriateness of the conditions which define what 'fits' and what does not. Only affective valuation puts an end to questions. But that is because it is not made from an evidential base. It may not be 'made' at all. It may just happen. Any valuation made from evidence is in danger of infinite regress due to the fact that the evidence can be questioned. But a valuation not made from an evidential base is in danger of being purely arbitrary—such that no sense can be given to its being correct or incorrect, for example. It is clear that finding a ground for valuations will entail dealing with the 'unfinished' nature of some acts of valuation and the apparent arbitrariness of others.

These questions will be dealt with in the course of outlining *e*valuation procedures in Chapter VI. Specifically, it will be argued that since second order questions, in the case of four of the types of valuation, only lead to third order questions, and thence to fourth-order, etc., what is needed is a way of putting a non-arbitrary stop to the process at some point. The object of such a grounding procedure will not be to stop all reason-giving at the level of second-order questions, but merely to provide a stopping point at some manageably finite 'order.' It will turn out that it will often be an advantage to one in defending a valuation to shift the focus of attention to questions of the second or third order before grounding the matter. For example: I take the medicine because it relieves my discomfort. (First order valuation: the medicine is 'good for' relieving discomfort.) I want my discomfort relieved so that I can work in peace. (Second order valuation.) I want to work in peace so I can work well. (Third order.) I want to do my work well because I can't be happy otherwise. (Fourth order.) I want to be happy. . . Here, perhaps, one will find it possible to *ground* the process.

4 FURTHER REMARKS ON THE MEANINGS OF VALUE-EXPRESSIONS

The typology of valuational acts introduced in Chapter IV makes it possible to see apparently conflicting value practices as in fact based on (and legitimized by) different types of valuational acts. The frequent insistence that a valuation's 'validity' cannot, even in principle, be challenged, and the equally frequent insistence that one can make mistakes in valuing things—that there is evidence that is relevant to the case—can be seen as proceeding from affective valuations on the one hand and summary, calculative, or good-of-its-kind valuations on the other. It is obviously true that in a case where the value-sentence expresses only the awareness of an affective response (with, perhaps, an implicit commendation) there is no question of the truth or falsity of the judgment. Mere *expressions* of feeling or other affective responses are not true or false any more than the feelings themselves are true or false. When and in so far as the question, 'What is the value of x?' is answered with an expression of a pro or con affective response, the only disputable points are (1) whether what was meant to be expressed was expressed (or vice versa) and whether it was well expressed (good of its kind?); and (2) whether the response expressed was appropriate, correct, right. The former question has to do only with getting the right words in use and thus is concerned only with the technique of expression, not the valuation itself; the latter is the evaluative question, to be dealt with later. So affective valuation is indeed indisputable in a clear sense.[1]

On the other hand, when a value sentence expresses only a calculative valuation (or, shall we say, when the expression is intended to convey only the result of the calculative element of that form of valuation, regardless of whatever other elements—imperative, commendational, affective—there may be in the total complex of the valuational act), then there obviously is a question as to the truth or falsity of the expression. There is also, of course, a question of the propriety (correctness, rightness) of the calculation—a question, again, of the evaluation of the valuational act. So the typological approach can give an account of both common sense platitudes: that there is no disputing taste; and that one can be wrong about a thing's value. But the 'essentialist' approach dies hard. Some further remarks, in terms of the problem of explaining the 'use' of evidence in value arguments are surely not superfluous.

One grants that a case can be made for the necessary presence of a pro/con affective response in all acts of valuation.[2] But to argue that these ubiquitous features 'are' what is 'essentially' valuational about the acts is, as has already been argued, misleading. To put the point again, and somewhat differently, consider that one feature of our speech, i.e., at least in our verbal attempts to express our experience, is that of emphasis. We leave out as much as we get in, the poet's skill notwithstanding, and in many cases this selectivity is precisely what we want. We want to focus attention, our own or others', on some special features of the experience. And so, to insist that value expressions must aim at expressing the necessary features of all valuational acts encourages a serious misdirection. For it assures us that we know precisely how to handle all value expressions once we get hold of the necessary features of valuing (since we know of what sort they all 'must be') and gives us confidence to push them all into the same pattern whether the language users protest or not. The typology, on the other hand, warns that not only is it important to pay attention to the whole of each valuational act (as opposed to paying attention to just the features they have in common), but that value statements may not always be designed at all to express the common, 'necessary' features of the acts, but may rather be designed to call attention to features special to each type of act. If so, then the 'logic of value sentences' is likely to vary from case to case, and what a value sentence means—whether it is an assertion or not, whether it can be evidenced or not—will have to be decided from case to case.

It is on the question of the relevance of evidence (reasoned argument) to valuations that this point becomes clearest. If one fixes on the 'essential' features of valuation—which seem, most plausibly, to be related exclusively to its various non-cognitive features (attitude orientation, for example)—then the use of evidence and reasoned argument is at best regarded as highly indirect.

For example, one might maintain that all differences or most significant differences of attitude are the product of differences of belief, since one's beliefs 'cause' the attitudes and there is a high degree of uniformity across the species on which beliefs cause which attitudes. Thus to find a difference of attitude (valuation) is to find a concealed difference of belief. Deal with the difference of belief in a reasoned way; get agreement if you can, and if you can, you will most likely get agreement of attitude. As the night follows the day . . .

But this view of things is properly resisted if, again, only because it insists upon a suspiciously partial analysis of what it means to value something. From the admission that the pro/con affective response is a necessary feature of all valuational acts and is sufficient to constitute a valuation, it does not quite follow that all valuation is properly described only, or even primarily, as constituted by such responses. And the impressive resistance offered, both by philosophers and others, to this view of the solely 'psychological' relation of evidence to arguments on value questions can only confirm the suspicion that the identification of valuing with this one (necessary?) feature of it has been hasty. Insisting that the 'whole' of each valuational act be taken into account, then, and not merely the features common to all, permits the conclusion that some valuations legitimize the use of evidence, are true or false, and others do not.

In short, the analysis offered in Chapter IV provides a more direct and satisfactory account of the common value practices mentioned in Chapter III than an essentialist account is likely to do.

NOTES

1 The indisputability of expressions *of* affective valuations is confused by the similarity they bear, in grammar and choice of words, to expressions *about* affective valuations which, of course, are usually assertions (i.e., true-false sentences). Expressions of the results of summary, calculative, and good-of-its-kind valuations are or contain assertions. The result of these activities is, after all, at least in part a *belief* about something; expressions of belief can be true or false. But the value discourse for affective valuations is not composed of assertions; only the meta-value discourse about affective valuations can contain assertions. But that meta-value discourse can be used in the process of evaluation (discussed in Chapters VI to VIII) quite straightforwardly. More discussion of this point later, but see, for a corroborative account, D. S. Miller, 'Moral Truth,' *Philosophical Studies*, 1 (1950), pp. 40–6.

2 It might be noted, by way of explaining the persuasiveness of the view that valuation is essentially or exclusively affective, that when one considers some experiences agreed to be non-valuational and asks what change in circumstances would make them valuational, he often finds that the necessary change is the addition of a pro/con affect. He then notes that the mere existence of a pro/con response can be sufficient grounds for calling an experience a valuation, and he is encouraged to identify valuing with such affects. But such an identification is hasty and does not (deductively) follow from the admission that the pro/con response is a necessary and (occasionally) a sufficient condition of an act's being a valuation.

VI

EVALUATION

The existence of the acts described in preceding chapters as valuations cannot reasonably be denied. One may deny that they are in fact valuations; arguments have already been given to meet such denials. But whether one agrees to the typology or not, he is still faced with explaining the significance of the acts of valuation he does recognize. One admits that people 'value' things, but so what? When a valuation has been completed, what conclusions may one draw, what has he learned, what help, if any, has been given to attempts to discover what he ought to do? It was noted that even if one grants the valuation—i.e., grants the truth (or a truth about) the value expression—one is still faced with second-order questions: e.g., the apparent arbitrariness of some, the apparent indefinite regress generated by others. But in addition to these questions, there are others arising from the fact that acts of valuation leave one with a welter of value expressions, many of them conflicting, all competing for attention. How is one to handle this chaos? To judge what *the* value of a given object is? And once that is decided, how is one to come to conclusions about conduct from conclusions about values?

Evaluation, as used here, refers to the process of sorting out, from the welter of valuations, those which may be called 'well-formed,' ranking the resulting values in terms of their relative weight, and balancing goods against the bads to come to a conclusion about whether the object is good (or bad) on the whole. It is at this point that most axiological schemes run into the most insistent trouble from scepticism; and it is at this point that the most careful attention to procedure is necessary.[1]

I PRELIMINARIES

To begin with, it will be assumed by what follows that when the truth of the value expression[2] has been granted, one has by that fact granted that the object valued 'has' or 'is of' the expressed value. In most cases these locutions are merely convenient shorthand for a much more complicated story—e.g., 'The hammer is a good one' expressing a rather involved set of relations between physical objects, properties of those objects, and human purposes. In some cases, of course, it is asserted that to say an object 'has' a value is to say something literal and only about that object. Whether one wants to say that the talk of an object's pleasure-giving potential is of this literally 'objective' kind, or whether one wants to reserve the label 'objective' for other views (e.g., Moore's in *Principia Ethica*), is another issue. For present purposes, the matter of subjective versus objective values is of no importance because procedurally, for the justification question, they are treated alike. A value is taken to be the 'fact' indicated by a true value expression. What 'fact' is indicated will depend on the nature of the act of valuation. Preceding arguments have described five types of valuational act. Value expressions associated with each type state the results of the acts, and the 'facts' indicated by each classify, one supposes, into five types of value. But care must be taken with this progressive objectification of value, and one precaution which will be taken at the outset is that no distinction will be forced between values *of* and values *for* an object. This is simply another way of putting the objective-subjective distinction, and as it will make no procedural difference from this point on, it will not be used.

What sense, then, does it make to talk about the value of an object? If an object can be valued in all five ways, even if all 'agree,' the object will still have five values. And since each object is likely to be valued affectively in more than one way (liked by some, disliked by others), and further, since each object is necessarily bad for as many purposes as it is good (just taking the negation of each 'good-for' purpose), it has even more values.

Consequently it is hard to imagine an object for which *the* value, as opposed to values, can be specified.

This result is not especially troubling until one comes to the question of decisions on courses of conduct. But then one does need a univocal conclusion if he expects to decide what to do on the basis of valuations. It is assumed that a decision based on valuations will aim at finding the best or the right course of conduct. And to do that, some means must be gotten for coming up with univocal valuations for pieces of conduct out of the array of values with which one is initially presented.

What are needed are some non-arbitrary ways of both narrowing the range of valuations which must be considered (so that one is not obligated to consider every conceivable value that might be placed on a piece of conduct), as well as some non-arbitrary ways of ranking values in terms of importance and of 'balancing out' goods and bads. These procedures, as noted, will be termed *e*valuative ones.

2 THE ETHICAL NEUTRALISM OF THESE PROCEDURES

On the matter of narrowing the range of valuations, egoism certainly appears, at least initially, to fill the bill, asserting, as it does, that unless one has compelling reasons to the contrary, he need only consider his own valuations—or at least, his own good. The reasoning behind this begins, one supposes, with the contention that things have value for me, and for you, and for. . . ., and each must act on his own values. But this is surely a misleading way of putting the findings of the present study. *People* value things, and consequently those things have values. My valuations can have, *prima facie*, no special status in the determination of a univocal value for an object, whether that object is a piece of my conduct or something else. All valuations, by whomever made, are *prima facie* on an equal footing. That is, to say that things have values when people value them is not (yet) to assert any priorities among the values. There are as yet no reasons for concluding that my values for my act are to be given more weight in a reasoned evaluation of my conduct than anyone else's values for that conduct. So there is no egocentric predicament here, no problem of justifying the inclusion of others' valuations in one's assessments. All valuations are included unless one can find reasons for excluding them. The burden of proof rests with the egoist.

And in any case, egoism usually turns out to be only an illusory way of narrowing the scope of one's considerations, for it is usually elaborated to get to all the conventional maxims about good deeds by way of figuring in the effects of others' acts (valuings, *et al.*) on one's own affairs. While egoism is certainly to be distinguished from the neutralist and altruist positions (even though many of the conclusions coincide), there is little if any difference among them in terms of the complexity of the procedures made necessary by the number and variety of valuations to be considered.[3]

3 WELL-FORMED VALUATIONS

What does work remarkably well in reducing the bewildering array of valuations is simply the straightforwardly justifiable principle that any value expression which can be shown false is to be ruled out. This needs little comment. Such expressions indicate no 'values' (if 'value' is taken to mean the fact expressed by true value expressions); they either entail anything (as in the usual interpretation of propositional logic), in which case they prove nothing; or they contribute nothing one way or the other toward determining the truth or falsity of a value expression when it is treated as a conclusion in a valid inference.

But there is another test for well-formedness which is just as effective at ruling out value expressions. To *confound* a valuation is to show that, for reasons of lack of clarity or lack of conclusiveness, the value expression cannot reasonably enter into the final assessment. An act of valuation which can be shown to be inconclusive (confounded) is strictly not a valuation at all, only an attempt at one. As such, the verbal expression is surely not the expression of a value; thus it can not enter into further considerations taking values as data, e.g., the 'balancing' of values—evaluation.[4]

Each type of valuation has its own special difficulties. Some of these were noted in connection with justifying value expressions from each. Summaries can be improperly made (not 'known to be false,' for that is another issue). They can be made in such a way that the results are inconclusive. This happens when something is wrong with the sample, for instance. To exhibit this inconclusiveness is to confound the situation—one cannot reasonably use an utterly inconclusive value expression further, for to the degree that it is not known whether the expression is true, it is not known what value the object has. And to the degree that one has no information on the object's value, he can hardly proceed to 'balance' its values. Similarly with calculative valuations which can be shown to be inconclusive: a situation, for example, in which for every indication that *x* is good for *y*, there is a countervailing indication that it is bad for the same *y*. Further use here is unwarranted. And the same is true of good-of-its-kind valuations where no type description can be agreed on and it thus cannot be agreed what kind of thing the object to be valued is. Or even if that can be agreed, occasionally no conclusion can be gotten on what counts as a deficiency or 'virtue' of the type involved. Or if that too can be decided, the valuation can still be confounded by exhibiting exactly countervailing deficiencies and virtues. Even affective valuations can be confounded by showing that the response involved cannot reasonably be called either a pro or a con one. In all these cases, the valuations (i.e., the use of the corresponding value expressions in further arguments about what course of conduct to take) must be ruled out. Though a man may try, in making his valuations, to get ones which cannot be falsified or confounded, it is not always and perhaps not often possible to reach an assurance that he has seen all the angles. This openness is in many cases precisely the inevitable openness of inductive argument, and contributes, of course, to the so-called 'looseness' of moral arguments.

In passing one might note that in certain cases, governed by the same slippery rules used for all legitimate appeals to authority and the need for expertise, one may want to disallow valuations merely from the fact that they were made under conditions which render their accuracy suspect—especially where one is faced with a choice between conflicting value expressions, one of which expresses a valuation which was *not* made under disadvantageous circumstances, and when there seems to be no direct way to determine the truth of either. In such circumstances, one would presumably have reason to opt for the 'undisadvantaged' one.

To say this is no more than to recognize that there are circumstances under which it is very difficult to carry out valuations of one type or another without getting a distorted result, and that when one is deciding the probability of a value expression's being true (or, in the case of affective valuation, considering the question of whether or not the response is hallucinatory in nature) the 'situation' of the valuer is an important consideration. This is the rationale behind the requirement for disinterested adjudication, pleas for 'informed' judgment in aesthetic matters, and arguments for the need for expertise in deciding the usefulness, effectiveness, and so on of sophisticated technological systems.

In the case of affective valuation, one often is faced with rather insistent valuations from people who have no significant experience of, and no adequate imagination of, the thing being valued. Their response is thus to an illusory object (their vacuous or misconceived view of the object they suppose themselves to be responding to) and thus cannot be accepted as a valuation of the supposed object, but only of the 'illusory' one. For example, one's revulsion at a mistaken conception of what it is to perform an autopsy must not be taken as a valuation of performing an autopsy. It is a valuation of another act altogether, which has been imagined to be (like) an autopsy.

Thus one can reasonably strike from further consideration all valuations made from circumstances which *guarantee* a distorted result, and can occasionally disallow valuations made from 'disadvantageous' circumstances. Again, however, such rulings are usually disputable and another contribution toward the looseness of justifications of moral judgments.

When one has gone as far as he can toward narrowing the range of valuations of the conduct he is considering, by ruling out false, confounded, and 'disadvantaged' valuations, he will still most likely be faced with a helter-skelter of valuations, many of which may be in conflict. And even if the remaining valuations are not in conflict, one still has to deal with the second order questions they raise (i.e., infinite regress versus arbitrariness). One needs to secure a means of ranking values in terms of relative importance—so that it can be decided, for example, in what sorts of cases calculative valuations ('The medicine is good for you.') 'outweigh' affective ones ('But I can't stand it!'), and which of two calculations for the same object (x is good for y, but bad for z) is to be decisive. Further, one needs to secure a base from which, and in terms of which, valuations can be justified in a way that does not generate either an infinite regress of justifications, or an arbitrary cutting off of the demand for justification.

4 THREE PROCEDURAL PRINCIPLES

To summarize the procedural principles for evaluation developed so far, it has been argued that:

(1) *Things have values when people value them.* Whether there are 'objective' values or not, there still are, and are likely always to be, valuations—values *for* things. And these will have to be dealt with in moral philosophy regardless of the outcome of the 'objectivist' controversy.

(2) *No one's values are* prima facie *in a privileged position.* Egoism is to be rejected by noticing that, as a consequence of (1), any person's valuation must be entered into the balance.

(3) *Any value-expression which can be shown to be false, can be confounded, or shown to be 'disadvantaged' may be ruled out as ill-formed.* The need now is for some 'ranking' and 'balancing' procedures which will allow one to decide, from the welter of valuations remaining, whether the object valued is good or bad on the whole.

NOTES

1 On this matter of confusing what are here called the valuative and evaluative levels, see the instructive and frustrating arguments between Dewey and his critics on a related matter. Dewey wanted to say that values resulted from appraising (an instrumental act). He distinguished appraising from prizing (the data of appraisals) and never argued for an instrumentalist or cognitivist account of the latter. His critics have a hard time understanding this, largely because they want to describe as values only the products of prizing and Dewey insists on reserving the word for the results of appraising. See the Dewey versus Philip Blair Rice controversy in articles in the *Journal of Philosophy* for 1943 and 1944.

2 For the sake of convenience, the more cumbersome phrase 'the truth of, or relevant truth about, the value expression' will be omitted. The expression of affective valuations is, of course, the cause of the trouble. One's expression of an affective response is neither true nor false. So for our purposes, we take the related truth *about* the expression in each case—namely that '*x* has *y*-

response to *z*.' But as long as this nicety is remembered, it will do no harm to shorten the phrase to 'the truth of value expressions.'

3 For some of the literature on egoism, see A. Edel, 'Two Traditions in the Refutation of Egoism,' *Journal of Philosophy*, 34 (1937), pp. 617–28; W. D. Glasgow, 'The Contradiction in Ethical Egoism,' *Philosophical Studies*, 19 (1968), pp. 81–5; Jonathan Harrison, 'Self Interest and Duty,' *Australasian Journal of Philosophy*, 31 (1953), pp. 22–9; John Hospers, 'Baier and Melden on Ethical Egoism,' *Philosophical Studies*, 12 (1961), pp. 10–16; Jan Narveson, 'Refutation of Egoism,' pages 268–71 in his book *Morality and Utility* (Baltimore: Johns Hopkins Press, 1967); Robert G. Olson, *The Morality of Self-Interest* (New York: Harbinger Books, Harcourt Brace & World, 1965); and the four articles defending hedonic egoism by Gardiner Williams in the *Journal of Philosophy*: 'Normative Naturalistic Ethics,' 47 (1950), pp. 324–30; 'Hedonism, Conflict, and Cruelty,' 47 (1950), pp. 649–56; 'Universalistic Hedonism versus Hedonic Individual Relativism,' 52 (1955), pp. 72–7; and 'Hedonic Individual Ethical Relativism,' 55 (1958), pp. 143–53. See also, on a much different and more elaborate plan for refuting egoism, Thomas Nagel, *The Possibility of Altruism* (Oxford: Clarendon Press, 1970).

4 For a somewhat different slant on some of the same matters, see Charles Baylis, 'The Confirmation of Value Judgments,' *Philosophical Review*, 61 (1952), pp. 50–8.

VII

GROUNDING VALUE JUDGMENTS

A physics which takes rest as a primitive concept and sets out to account for the existence of any motion whatever has rather different problems from one which takes both rest and motion as primitive and tries to account for changes in either state. There is a thinly analogous difference in ethics between theories which demand a justification for any norm and those which grant the existence of some and require justification only for change. This chapter and the next argue for a version of the latter sort of ethical theory by treating some norms *presumptively*.[1] The 'openness' obtained by this procedure is intended to rebut the usual objections to natural norm theories. This chapter will, after some necessary preliminaries, describe three presumptive value criteria and answer a few initial objections against their use. Chapter VIII will describe in more detail the way the criteria may be used to ground the justification of value judgments.

It should be noted, however, that it is the validity of the *procedure* which is primarily in question here—not the adequacy of the three criteria used as examples of the procedure. This accounts for the rather unremarkable analysis of the examples. If one were to develop a substantive moral theory along these lines, a sophisticated analysis of each criterion would be required. But even then one could not reasonably suppose he had anything more than a special-purpose model of human nature.

1 A REPRISE OF THE SCEPTIC'S ARGUMENT

It will be recalled that on the matter of a rational basis for morality, scepticism proposes a dilemma something like this: any attempt to give a thoroughgoing, reasoned justification of one's fundamental moral principles[2] (or values, or obligations, or character assessments) will eventually come to a point at which one will either have to put an arbitrary stop to the demand for justifications (in which case he has given up the attempt to get a *thoroughgoing* reasoned justification), or one will have to go on indefinitely—giving reasons for principles, reasons for the reasons, reasons for *those* reasons, and so on.

Such scepticism has been applied to some theories (e.g., intuitionism) with the contention that their procedures are empirically indistinguishable from putting an arbitrary stop to the reason-giving process. It has been applied to others (e.g., sophisticated forms of naturalism such as Dewey's, and the theories of the 'good reasons' school) with the contention that they only avoid an indefinite regress by involving themselves in logical circles: justifying, for example, specific policies or principles for minimizing value conflict in terms of general features of rationality *per se*, or morality *per se*, and then justifying the adoption of the rational mode of being or 'the moral point of view' in terms of the need for minimizing value conflict.

The usual rebuttals to the sceptic's contentions—in addition to frequent expressions of impatience—are either to the point that all knowledge is ultimately based on 'assumptions' (and thus that moral knowledge is in no difficulty peculiar to it), or they are to the point that some logical circles are not vicious ones after all, and that if the basis for morality is no more than rationality or 'reasonableness' *per se*, then the demand for a reasoned justification of this basis is itself circular—a demand for reasons for reasoning. Further, as an elaboration to such rebuttals of the sceptic's point, the use of a straightforward deductive model for ethical reasoning has occasionally been criticized as misrepresentative of our actual procedures in moral reasoning and as unsuited to our actual needs.[3]

These rebuttals have raised storms of protest and reply, very complicated secondary issues, and ingenious attempts at synthesis and/or compromise among competing positions. Surely one motive for a shift in focus is to get some distance on the current maze of argument, counter-argument, adjustment, accommodation and renewed attack.

The tactic proposed here is simpler, and, I think, more effective than others in shifting the burden of proof to the sceptic. Once the sceptic, rather than the moralist, is under interrogation, the possibility of a rational basis for morality does not seem so remote.[4]

2 THE DIFFERENCE BETWEEN A GROUND AND A PROOF

The essence of the argument to be made here depends on a distinction between *grounding* a judgment, defined as putting a non-arbitrary stop to reason-giving, and *proving* a judgment, which is an offer of reasons. The distinction between a ground and a proof, in turn, depends on noticing that not all 'rational starting points' for arguments are either assumptions, in the usual sense, or conclusions from prior evidence.

Consider an analogy. There are some logical principles, for example the old 'laws of thought,' which are not assumptions in the sense of being propositions whose truth one actively assumes (outside the context of certain metaphysical inquiries). And they are not propositions for which one ordinarily offers proofs, either. Such principles are 'built-ins'—parts of the normally-formed human's conceptual apparatus. They can be challenged, and thus are not to be confused with 'self-evident truths,' but they are in no ordinary sense chosen, adopted, or assumed.

If no reasons can be offered for abandoning such principles, then they may be defended as *grounded*, though 'unproved.' For if a principle has not been assumed or adopted in the first place, it cannot (*a fortiori*) have been *arbitrarily* assumed. And if there is no reasoned argument *against* it, then it cannot be, from a reasoned point of view, wrong. At worst it is a matter of indifference to reason. This line of argument is cleaner and quicker (though perhaps less exciting) than trying to argue that the principles involved are necessary components of 'being rational' or 'being human,' and it is more effective for present purposes.[5]

How this sort of argument can be turned to the moralist's

advantage is fairly obvious. One takes care to found his moral theory on judgments which express 'built-ins'—norms or values or imperatives which are parts of the equipment all normally-formed human beings have—and asks for some reason to *refuse* to use them. One uses these judgments, then, as *presumptively* valid, meaning merely that the burden of proof lies with the sceptic. And this is not a trivial or arbitrary turning of tables, as it may seem to be at first glance.

3 PRESUMPTIVE VALUE CRITERIA

As an illustration, I want to describe several features of human life which can reasonably be taken as 'givens' (i.e., features of our lives which emerge without our having chosen them, and which do so in every normally-formed human being), and which may be used as *presumptive value criteria*. Some remarks on terminology may be in order.

'Normally formed' here means statistically normal among the offspring of human beings. No doubt some abnormal human offspring may be found in which the features of life to be described do not occur. But excluding them from consideration at this point seems reasonable if only because (as will be obvious when the features are described) one could hardly imagine a creature who lacked them being a 'valuer' in any sense at all.

'Value criterion' here means any criterion which sorts values according to relative importance, thus assigning priorities, or 'weights,' in the process of balancing goods and bads to decide the best or right course of conduct.

'Presumptive criterion' here has a meaning adapted from legal jargon. A 'legal presumption of innocence' means merely that the burden of proof rests on the accuser, not the accused. To say that a man is legally presumed to be innocent is not to say that necessarily someone *presumes* (in the ordinary sense of having a conscious conviction) that he is innocent. It simply means that in the course of carrying out the law, if no evidence is found which is sufficient to show his guilt, the accused will be *adjudged* innocent. Similarly, here, I shall argue that in the quest for a reasoned justification of value judgments, some criteria may reasonably function *presumptively*—that is, that they may be used non-arbitrarily as value

criteria as long as no sufficient reasons to the contrary are found.
For example:

PURPOSIVENESS

Human beings, within the range called 'normally formed,' are purposive as long as they are alive, awake, and do not choose to change. And of course purposiveness is not something one chooses, for choosing is itself a purposive act. We simply become purposers in the normal course of events, and, as long as we remain alive and awake, we simply *remain* purposive unless and until we take ourselves out of the business, for example, by suicide or drugs.

(a) A purpose operatively defines a direction for conduct and has an operational criterion of success, the 'satisfaction' or 'fulfillment' of the purpose. (b) The awareness of the plurality of purposes, the conflicts among them, and the impossibility of satisfying all the purposes that might reasonably occur in one's life also defines operatively a direction: the satisfaction of as many as possible. (c) Further, that awareness defines (again operatively, if not explicitly and propositionally) priorities *among* purposes: one opts for the purposes and satisfactions that are pervasive through time and across the breadth of one's concerns; one opts for the fecund, something productive of more of its kind, over the sterile; one opts for the long-lived over the short-lived, the intense over the mild. (d) Further still, as awareness of the enormous complexity of possible purposes grows, along with awareness of the mechanics of getting the most satisfactions for one's efforts, one will, when faced with a conflict, be inclined to opt for satisfaction of pervasive and fecund purposes over, for example, immediately compelling but fleeting ones.[6]

It should be explained that when it is said that value criteria define priorities operatively, if not propositionally, it is only meant that very often these features of life operate to establish priorities in one's conduct without one's 'propositionally formulated' awareness of what is going on. But for purposes of reasoned justification procedure, one produces the formulations—just as one formulates the results of affective valuations into assertions—so that they may enter, logically, into reasoned arguments.

Evidence for the emergence of these operative priorities as

a normal consequence of development is provided by looking at the way in which one is taught about 'delayed gratification,' the value of learning to read, taking medicine, etc. When such things are taught, the procedure is usually simply to call the learner's attention to the complexities and conflicts and mechanics of purpose-satisfaction. The assumption is that one will simply 'see' the need for the priorities mentioned.

It must be remembered that one is not here arguing for the greater *value* of one sort of purpose over another. That route would indeed lead to a demand for the justification of the 'greatness' of its value, and presumably to either an infinite regress or an arbitrary cutoff of justifications. What is argued here is rather that these priorities are already in operation for all purposive humans; that they emerge unchosen in the normal course of events for normally formed humans; and that they make it possible to use the notion of purposiveness as a presumptive value-criterion.

What is meant by 'presumptive value-criterion,' again, is simply that the direction and priorities operative from the given purposiveness of our lives can reasonably be allowed to function as a value criterion as long as no reason to the contrary can be justified —that there can be no reasoned objection to proceeding along a given line if no reason for changing the line can be secured. When there is no reasoned objection to an existing set of priorities, there is no further 'need' for justification of those priorities; and if those priorities were not themselves chosen, but simply happened, there is nothing 'arbitrary' about them.

Now of course the priorities given by purposiveness can be confounded by conflicts—if, for example, one has to choose between a pervasive but sterile purpose and a fecund but fleeting one. In such cases, it may be that presumptive guidance is *not* given from the nature of purposiveness, and that unless some can be gotten from the other presumptive value criteria (see below) or *ex nihilo*, no reasoned decision will be possible. Such cases of course add to the notable 'looseness' of moral argument.

PERSONALNESS

Human beings, within the range called 'normally formed,' are selves as long as they are alive, awake, and do not choose to change. This awareness of the world as divided into self/other is again not something one can choose to get, for 'choosing to get' an awareness of the world presupposes the awareness of the distinction. The emergence of personhood simply occurs in the normal course of events, and though we may choose to take ourselves out of it, by suicide for example, we remain at least dispositionally aware of ourselves as persons as long as we are alive and awake and not so acting as to exclude the awareness.

An awareness of personhood includes an operative definition, vague though it may be, of boundaries, an incursion into which *under certain circumstances* is taken as a *prima facie* offence and generative of a defensive response. Whether this defense is aggressive or passive, and under what conditions it is likely to be one or the other are not important for procedural purposes, though this is indeed an important issue for moral psychology. Just what sorts of 'incursions' will generate a defensive, 'boundary-keeping' response and which (like invited intimacies) do not, is a matter of the first importance for moral psychology. But the crucial item for the discussion here is simply that people do, operatively, discriminate among 'incursions on their person,' accepting, enjoying and even encouraging some, while rejecting others. Further, it may be argued that one's awareness of his own intrusion into (violation of) another's personhood is typically accompanied by fear, aversion, perhaps incipient guilt, though the last is usually taken to be a more complex matter.

Empirical psychology has just begun to get a grasp on these matters,[7] and it would not be appropriate to guess at the possible findings here. But it is possible to see that the awareness of personhood can provide presumptive value criteria in every bit as strong a way as the purposiveness of our lives does. The existence of operative priorities for conduct in response to a violation of one's person is taken as given and unchosen. If no reason can be secured for changing these priorities, there can be no reasoned objection to going ahead with them. If there is no reasoned objection to a course of conduct, there is no further 'need' for justification. And if these priorities were not themselves chosen, but simply happened, there is nothing 'arbitrary' about them.

One can of course confound the priorities here by showing either that a supposed intrusion is not, or at least not clearly, an intrusion, or that the intrusion is not, or at least not clearly, an offence. Similarly for whatever operative priorities may be attached to one's awareness of violating the person of another. When the priorities are thus confounded, no presumptive guidance is given by the nature of 'personalness,' and unless some can be gotten from other presumptive value criteria, or *ex nihilo*, a reasoned decision on what to do may not be possible.

THE AESTHETIC NATURE OF LIFE

Human beings, within the range called 'normally formed,' are seekers of pleasure and equilibrium, avoiders of pain and disequilibrium as long as they are alive, even semi-conscious, and do not choose to change. This disposition is prior to and much more subtle and pervasive than conscious, purposive pleasure-seeking. It can be noticed to some extent in infants, the sleeping, the senile, and even occasionally the comatose. It may be distinguished from the homeostatic mechanism at the cellular level, of course, and from the analogous mechanisms of the autonomic nervous system. It operates on a 'grosser' level than these others (which is not a claim for its causal independence from them, only a descriptive distinction). It is at least an adjustment mechanism, perhaps a propensity or an even more self-initiated thing, unchosen by us, which is present as long as we live, are responsive to gross sense stimuli, and do not act to eradicate it or change it.

The aesthetic disposition defines a set of operating priorities in terms of readiness to respond toward or away from, with or against, mildly or violently to different situations. These readinesses may be used as presumptive value criteria, like purposiveness and personhood, features of our lives which do in fact sort options, valued or not, into a rank order and which one can reasonably allow to operate to sort his valuations in a similar way as long as no reason to the contrary can be secured.

4 SOME LOOSELY METAPHORICAL COMMENTS ON THE USE OF THE CRITERIA

The operation of these three sets of dispositions with regard to sorting valuations of courses of conduct depends, of course, on the conceptual ability to 'connect' the course of conduct with the dispositions in such a way as to activate the readiness to respond. The response mechanism operates in respect to existing situations. Its use as a presumptive value-criterion demands that it operate *pro*spectively (before the fact, in deciding what to do) and *retro*spectively (after the fact, in assessing liability for blame for example). This obviously involves further conceptualization—perhaps 'confronting oneself' with an imaginative 'event' in order to 'feel' the sorting operation take effect. But an adequate account of such metaphors is beyond the scope of this work.

Indeed, even an adequate description of the three criteria is beyond the scope of the present discussion, due to the complexity and deep-dispositional nature of the features being described. The problem of description here seems similar to those faced by writers of generative grammars where the attempt is to describe a similarly dispositional, complex, and potent human ability in terms of sets of rules. When such discursive descriptions are attempted for purposiveness, personalness, and the aesthetic disposition, as they must be if we are to use these features of life as criteria in reasoned argument, the result inevitably seems pathetically static and dry and for those reasons inadequate. Artists, generally, do a more adequate job of understanding and suggesting to others the vitality and force of such things than philosophers or psychologists. But the use of the dispositions as presumptive value criteria requires the discursive formulation, juiceless as it may be. In any case, one can still sketch, more or less clearly, the *procedures* involved in the use of the criteria. We shall have to be satisfied with that for the moment.

To confound the use of a presumptive value-criterion one attempts to interfere with the 'connection' between the course of conduct and the disposition. This may be done, conceivably, by justifying a change in the description of (one's understanding of) the course of conduct such that the 'connection' can no longer (clearly) be made. And this is, in fact, a very standard procedure in practice. When we are in fundamental disagreement with someone we often find that the *only* way to make progress is to argue about 'the *interpretation* of the facts'—i.e., how we are to describe our experience. Often a change in description produces a change in how one 'feels' about it—not only in how he values it, but in the priorities he attaches to those values in relation to other values. The difference here between propaganda and legitimate argument is made by whether or not the change in one's description, point of view, or understanding of the situation is a logically warranted change.

When the priorities of a disposition are confounded, it no longer can offer presumptive guidance on the ranking of valuations. And unless such guidance is forthcoming from one or more of the other presumptive value criteria, or can be gotten *ex nihilo*, again, no reasoned decision on what to do may be possible.

5 SOME OBJECTIONS

Suppose it is urged that this business about 'presumptive' value criteria is no more than a trick. And suppose one simply offers—as a reason for not using the criteria—that one does not want to use them. Will that not count as a 'reason' to the contrary?—destroying the presumptive validity involved? The answer to such an objection must be, of course, that this is no more a reason *against* the presumptive criteria than one's offering a statement of his wants in favor of some ultimate moral principle would be considered a reason *in support* of that principle. The point is that some reason against the presumption must be secured *in just the way* it has always been demanded that a reason in support of an ultimate moral principle be secured. The one demand is surely no less intelligible or warranted than the other. As for its being a trick, it is no trick to claim that if no reasoned objection can be offered to some course of action, that course of action 'needs' no further justification. *The failure to find reasons against doing x constitutes a failure to find x to be unjustified or wrong from a reasoned point of view. And that is enough to 'ground' the matter, to stop the demand for justification. For if x cannot be reasoned to be wrong, then at the very least it will be unobjectionable from that standpoint and the questions cease.*

Objection: 'But the turning of the tables here is trivial—a mere changing of the rules of the game. What philosophers have been after is not just "a way out," but a positive set of reasons. A proof. *That* is the challenge of the grounding problem, and these presumptive criteria do nothing to meet it. They simply refuse it.'

Reply: If philosophers have been aware, all along, of a 'way out' of the grounding problem which gives morality a rational basis in practice, then it is hard to understand their references to the spectre of relativism. Why worry about it if, knowing how to avoid it, you are merely trying to clinch matters by finding positive proofs for some fundamental moral principles? On the contrary, the crucial weakness of even highly sophisticated accounts of moral justification—ones which, for example, argue for the use of value-criteria rather than for some concept of 'higher' or ultimate value—has been that they do not use the value-criteria in an explicitly presumptive way. They thus open themselves unnecessarily to the charge of ultimate arbitrariness, or circularity, or infinite regress.[8] Surely the line of defense suggested here is not, then, a trivial one.

Objection: But surely we have many dispositions, in addition to these, which establish priorities. Some of these are learned and some are both *chosen* and learned, but they by no means just 'happen' in the normal course of events. Further, purposiveness and the others are incredibly refined and modified (due to cultural influences, choice, etc.) over the course of one's life. What about these?

Reply: While the use of a disposition as a presumptive value criterion does not depend on its being unlearned, it does depend on its being 'given' prior to choice. Otherwise, the question of arbitrariness in the choosing can be raised. Further, if one wants to use such criteria to help resolve cross-cultural value conflicts, he will need to use dispositions that are *species-wide* in order to start from ground common to the disputants. That is, he will need to use dispositions which are either inborn, or developed in the course of normal human maturation, or developed as a consequence of the *interaction* of such inborn or maturational dispositions with special (cultural or brute environmental) circumstances. That is why the examples of criteria here are so rudimentary.

It may be demonstrable, of course, that the priorities given by these rude features, when allowed to operate without unjustified interference, produce in the normal course of events more sophisticated 'second level' norms which can then be used in the same way. One thinks, here, of the way in which a child who is not bludgeoned into a psychosis develops the sense of being an 'agent' and comes to resist the pre-emption of his agency as a violation of his person. People who 'make our decisions for us,' order us around with naked and pointless displays of power, 'indoctrinate' us . . . all are resisted in this way. And perhaps an astute analysis of human development would show the 'normal' emergence of such resistance. If so, if it emerges from simply allowing the individual to develop within the frame of priorities given by 'primary' criteria, then it can also be used as presumptive value-criteria. Deciding which dispositions, and then which features of those dispositions, can properly be given this status is admittedly a difficult task. But this difficulty does not negate the *procedural* point that such criteria can provide a grounding of justifications.

Objection: Hasn't a virulent form of ethical egoism been introduced here? It looks as though the demand is that all valuations be rank ordered by the presumptive priorities given by one's *own* purposing, or awareness of personhood, or aesthetic disposition.

Reply: It has not been argued that one is to consider valuations by whomever made for his conduct in terms of his *own* personhood, etc. But that valuations by whomever made rank in importance by *that person*'s presumptive value-criteria and then enter into *the* final estimate, i.e., anyone's, of the balance of values. So a capricious desire to kill x is 'outweighed' by the fact that the act would definitively end the purposing of x; the act from x's point of view (getting killed) is the archetype of the 'sterile' purpose, to put it mildly. X's view of the act results in a value for the act (my act), a value which takes priority over the one resulting from my valuation in terms of the purposiveness criterion, due to the fact that a capricious purpose is to be subordinated to a fecund one, staying alive. No matter that the one purpose is 'mine' and the other x's. They each generate values for the same act, my killing x, and the capricious one loses in the balance.

Objection: But surely it is my act, and my purposes ought to have a special relevance to it?

Reply: Why? That is, they do if one can *show* that they do. Otherwise, one has no reason to conclude that they do.

Objection: If values exist when people value things, does this mean a man who has never 'valued' his life (that is, explicitly gone through a process you call valuation) is ignored when it comes to balancing off the values for and against his death?

Reply: I see no reason why the same sort of extensions one makes for the verb 'to know' cannot be extended to the verb 'to value.' That is, it is hardly reasonable to say that the only legitimate or important sense of 'to know' involves the knower's being able to formulate what he knows in propositions—or worse yet, to be attending to 'what he knows.' We do not cease to know, though we cease for a while to be *knowing* things, when we go to sleep. One does not fail to 'have' a value for one's life simply because he has failed to self-consciously go through the procedures called valuations.

Objection: But then one is quite at a loss to say who values what. Is the standard to be those who would (we feel certain) value the thing if they were asked? Or who would if they were to think about it seriously? Or who would if they *could* (e.g., the mentally defective)?

Reply: Again, the procedure with reference to who can be said to know something can reasonably be adopted here. And if this rules out entering any valuations from the mentally defective and from brutes into the balance, so be it. We have many other things to consider before we can say we have justified doing harm to them: considerations of the value of the practice or rule that would be involved in such conduct; the duties and obligations related to such an act which can be justified; and matters of moral character.

Objection: Still, it is not clear how one knows what another's operative priorities are. When one wants to apply these procedures to cases, how does he determine that *x*'s value is outweighed by *y*'s value? Determining that one valuation *x* makes outweighs another valuation that he makes is open to at least a rough determination in terms of a 'preference test.' But how does one lay one man's valuation alongside another's for comparison?

Reply: This is where the description of some species-wide priorities is crucial. For if such priorities are the ones at stake, one may reasonably argue that, unless there is some reason to think otherwise (e.g., one party is demonstrably not normally-formed), a simple thought-experiment about the relative weights of one's own values for the issue at stake, perhaps having to do with temporary pleasure and self-preservation, can be used as an indication of the way *x*'s temporary pleasure is to be ranked with respect to *y*'s self-preservation.

If the priorities involved in the case are not species-wide, or not demonstrably species-wide, one next asks if they are common to the parties involved. And of course if all this fails, as no doubt it often does, there are many other issues yet to be considered: e.g., the results of applying deontological and agent-morality considerations. The failure of an axiological scheme is not the failure of the justification enterprise *per se*.

6 SUMMATION

As a final thrust against the force of these arguments, I suppose one might try the tactic of sweeping denial. That is, one might argue that since no more importance is attached to these features than that they simply are, or happen, no more importance can be attached to the directions they define than to happenstance. What if humans change? What happens to 'presumptive guidance' then? Do all the old judgments change? Or do we suddenly decide that 'accepting' the change as a new value criterion would be 'bad'? If so, how do we decide this?

These are interesting questions and show the 'openness' gotten by treating these features as only *presumptive* criteria. There is, on this view, always the logical possibility that reasons may be found to support a change to other criteria, or to more. And the possibility of an evolutionary ethic is accommodated nicely. If a change is not 'made,' but simply 'happens' evolutionally, that too would change the presumptive value criteria. And one imagines that if humans were to change radically in the course of biological evolution, they might well need a radically different ethic.

In any case, the procedural point seems clear: one may use whatever species-wide priorities are 'built-in' to our natures as *presumptive value criteria*. And the grounding of justifications provided by such criteria defeats one of the sceptic's major objections to moral theory.

NOTES

1 Moral philosophy has recently borrowed a family of principles from jurisprudence and attempted to put them to work vindicating crucial moral precepts. The borrowed principles are those defining the notions of defeasibility, presumption, and burden of proof. The moral precepts to which they have been applied include those surrounding the notion of a 'free' act and those used to justify obligations and responsibilities. See H. L. A. Hart, 'The Ascription of Responsibility and Rights,' *Proceedings of the Aristotelian Society*, 49 (1948–9), pp. 171–94, for example. A natural extension of such applications has been carried out by Chaim Perelman in *Justice* (New York:

Random House, 1967). His attempt is to characterize a legitimate (just) law as one which is non-arbitrary, and to appeal to the practical principle that it is *change* (not all activity) which needs justification. The arguments in this chapter owe a great deal to these developments, particularly Perelman's work, though they are not, directly, a wholehearted use or a thorough critique of them.

2 For the attempt to ground morality on some indisputable principle, one thinks of Marcus Singer's work on the generalization argument, *Generalization in Ethics* (New York: Alfred A. Knopf, 1961). Arguments for a *summum bonum* can also be construed in this way, and of course Kant claims to be after the supreme principle of morality in the *Foundations of the Metaphysics of Morals* (trans. L. W. Beck, Library of Liberal Arts, 1959). For attempts to secure an indisputable *criterion* (which get closer to the view to be argued here), one might cite the recent work of Jan Narveson, *Morality and Utility* (Baltimore: Johns Hopkins Press, 1967) especially Chapter IX. Even more interesting (though less helpful) is the attempt to ground morality on 'transparent cases' —cases in which rightness or wrongness is taken to be obvious and in which further discussion is ludicrous: see William Gass, 'The Case of the Obliging Stranger,' *Philosophical Review*, 66 (1957), pp. 193–204.

3 For an alternative to the usual deductive model for moral reasoning, see the proposal by V. J. McGill to use negotiation as the model in 'Scientific Ethics and Negotiation,' *Proceedings and Addresses of the American Philosophical Association*, 42 (1968–9), pp. 5–20.

4 For a similarly oriented account of these matters, see A. Sesonske, *Value and Obligation: The Foundations of an Empirical Ethical Theory* (Berkeley: University of California Press, 1957).

5 See P. H. Nowell-Smith, *Ethics* (Penguin Books, 1954), Chapter 6, for a related point.

6 For a compelling account of related operative priorities, see John Rawls's *A Theory of Justice* (Cambridge: Harvard University Press, 1971), pp. 424–33.

7 The considerations advanced here about the operative definition of boundaries are a bastardization (and no doubt an unsatisfactory one from a psychologist's point of view) of some findings of both personality theory (the 'proprium,' personal distance, etc.) and experimental psychology (e.g., the work on fear/aggression). The details are, luckily, not as important as the outline in this procedural study. See for elaborations, Gordon Allport, *Becoming* (New Haven: Yale University Press, 1955); and Konrad Lorenz, *On Aggression* (Bantam Books, 1967).

8 For work which considers the grounding problem explicitly, but fails to consider using criteria presumptively, see: R. F. Creegan, 'Natural Law,' *Journal of Philosophy*, 43 (1946), pp. 124–32; Herbert Feigl, 'Validation and Vindication,' in Sellars and Hospers, *Readings in Ethical Theory* (New York: Appleton-Century-Crofts, 1952), pp. 667–80, especially page 675; Paul W. Kurtz, 'Need Reduction and Normal Value,' *Journal of Philosophy*, 55 (1958), pp. 555–68; Bernard Peach, 'Analysis and Criteriology in Philosophy of Ethics,' *Journal of Philosophy*, 52 (1955), pp. 561–71; and Paul W. Taylor, Chapter 5 of his book *Normative Discourse* (Englewood Cliffs: Prentice-Hall, 1961).

VIII

GROUNDING VALUE JUDGMENTS (CONTINUED)

It remains to be discussed how the presumptive criteria operate in the balancing of values. At least two problems loom large at the outset: in what sense can one 'balance' values, even if they have been rank-ordered by the presumptive criteria; and what happens when the priority-rankings of the same value by the various criteria conflict?

It should be remarked that no attempt has been made here to rank whole types of valuational act—e.g., to try to show that calculative valuations always outweigh conflicting affective ones, and so on. Rather, the procedure is to take each well-formed valuation, of whatever type, and put it directly 'up against' as many of the presumptive criteria as can give it a ranking. When all the available options have been valued and ranked, the methods introduced in this chapter attempt to decide which of the options can reasonably be judged the best one.

1 THE 'RIGHT' THING TO DO

First, some remarks on what conclusion one expects to get by 'balancing' the values. In the case of reaching decisions about conduct (to which, for convenience, the discussion here will be confined), one wants to discover the right thing to do—which must be defined, one supposes, as the best of the available options. Whether the options are predominantly good (i.e., good by some well-formed valuations and bad by no countervailing ones), or whether they are predominantly bad, the right course of action is the one productive of the most good, or the least bad, respectively. This is taken to be the meaning of saying that a course of action is the right thing to do.

It is true that when the procedures under discussion are applied, the range of conduct-options is highly circumscribed by the need for a decision by a more or less definite time, and by other circumstances which render the consideration of some options pointless. But it must not be supposed that the 'available' (i.e., possible of accomplishment) conduct options for a given situation are the only ones relevant for consideration. For it is very often useful (from the standpoint of finding the best possible course of conduct) to ask which of the available options most nearly approximates an ideal—or renders a future approximation more likely. What this consideration does is to provide a 'good-for' or good-of-its-kind valuation of available options in terms of an ideal (i.e., to provide another value for some of the options) and this may be useful in rendering a decision on what the best of the available conduct options is. Moral theory—as opposed to casuistry—may set for itself the task of securing potent ideals for this purpose, ideals whose function is as an auxiliary to the presumptive value criteria in effecting a decision in difficult cases. Such auxiliaries are fortunate, for, as has already been noted, the ways in which the valuational approach can be confounded are numerous. Any way, direct or indirect, of adding valuations into the balance increases the chance of breaking a deadlock.

It may not always be possible, however, to show that there is a 'best' course of conduct, and when we are convinced that such is the case, we say, 'There is no right thing to do,' or perhaps, 'There is no *one* right thing to do.' The latter locution implies that a distinction might be drawn between *a* right thing to do and *the* right thing to do. But it is a somewhat delicate distinction to make, for it does not seem to work at all for a set of equally bad options. If the available options are all predominantly and equally good, one can understand the impulse to say, 'Any one of them would be right to do,' implying that any course of conduct that is predominantly good is *a* right thing to do (the best of the lot being *the* right thing to do). But if the available options are predominantly and equally bad, one hesitates, it seems, to say, 'Any one would be right to do.' One will have to be done. And in terms of their justifiable values, any one will do. But none are 'really *right* to do.'

The intricacies of usage here will get further discussion in the discussion of the move from 'right' to 'ought.' But they will cause no difficulty for the time being if it is remembered that when no course of action can be shown to be the right one (the best of the available options), no reasoned decision solely and directly in terms of the value of the conduct is possible—i.e., the choice is from that point of view a matter of indifference. After all, when one faces an array of equal goods or equal bads, what more is there to say? There is certainly nothing more to say solely and directly in terms of the methods introduced so far. It will turn out that once the move from 'right' to 'ought' has been made, there will be something more, but the question for now is simply how one can show that a given course of conduct is or is not the best one of the available options.

2 APPLYING THE CRITERIA

In cases where only one of the presumptive value criteria is involved, the procedure is fairly straightforward. One first values the various conduct-options. That is, he finds out what well-formed valuations of each option can be secured. One does not wish to proceed with just whatever valuations happen to be at hand, but rather with all the valuations it is possible to get. This step in the procedure is forever disputable unless one can, in a given case, come up with a general method for exhausting the possibilities. But disputable as the step may be, it must not be considered *disputed*, and thus in need of defense, alteration, or reconsideration until it can be shown that some valuation has been overlooked. This is no more than standard procedure for dealing with a *ceteris paribus* clause in an argument, moral or not. One cannot always (or perhaps even very often) *prove* that there is no other evidence relevant to the conclusion, just as he cannot *prove* that there are no bats in the attic. But he can look hard enough to establish a presumption in favor of that contention, and that is enough for present purposes.

The demand that one secure all possible valuations of a given conduct option, rather than take only the ones people have in fact made, is necessary if one expects to conclude that the conduct in question is or is not *justifiable*. If one is content with the conclusion that the conduct has or has not been *justified*, the matter is different. In that event, all that need be included are the valuations actually made. But it is whether or not a piece of conduct is justifiable, surely, which interests the moral philosopher. He may have to settle for the other in some cases, but usually the conclusion about whether or not the conduct has actually been justified is of use more to adversary proceedings than philosophy (debates rather than arguments, trials rather than inquiries). The fact that it is often difficult to show that one has exhausted the list of possible valuations contributes, again, to the 'open texture' of moral arguments, but while it may render most procedures disputable in principle, it does not by itself—as was pointed out—render any of them disputed.

The second step is to rank the values for the conduct in terms of each value criterion. This is a notoriously difficult thing to describe, but as an activity, it is really quite commonplace. The ranking provided by the purposiveness, personalness, and aesthetic nature of our lives is simply *done* (not decided upon): the priorities are dispositional—readinesses to respond in certain ways. So to 'get' a ranking to use (presumptively) in a reasoned argument, one can only submit the options to his reflections and wait for them to fall into some order of preference.

What makes this step in the procedure disputable, of course, is the possibility that the preference one becomes aware of is given, not by any built-in features of our equipment as human beings, but by some overlay of conventional convictions or attitudes. But while such matters are difficult, they are not impossible in principle (for the nature of built-in preferences is amenable to empirical investigation), and it is often possible, in practice, to decide that one's preference for one thing over another is culture-relative.

So to repeat, the second step is to rank the values for the conduct-option in question. For example, if the conduct is good for *x* but bad for *y*, and *y* is presumptively a much more 'important' purpose than *x*, then the bad outweighs the good. The metaphor of weight is unfortunate here, for it hints at something more than an ordinal scale—e.g., an interval scale or a ratio scale. All that the presumptive value criteria provide is, of course, rank ordering.[1] But in many cases that is enough.

It may be helpful to schematize these procedures. Suppose two people are mulling over the value of a certain course of conduct in terms of the satisfactions it will bring (i.e., the sense of satisfaction in achievement of purpose). They consider, in turn, what other people seem to think about such conduct (valuation 1 on the schema below); how they 'feel' about it bringing them satisfaction (valuation 2); whatever hard evidence they can gather about the conduct's *effectiveness* in bringing about what they want (valuation 3); its value as something good of its kind (valuation 4); and its 'fittingness'—e.g., it is fitting conduct for people of their 'eminence' in society—(valuation 5).

Suppose one finds that people strongly approve while the other finds only mild approval. In terms of their affective responses, one strongly *dis*approves, the other is rather mildly inclined towards it. Calculatively, and in terms of its being something good of its kind, again (it need not be the same one in each case) one strongly disapproves while the other mildly approves. In terms of fittingness, they both value it negatively, though one rather less vehemently than the other. The rank ordering thus looks like this :

	Valuations	1	2	3	4	5
Rank	I	pro	con	con	con	con
	II	pro	pro	pro	pro	con

The balance is, in this case, obviously against the conduct, at least in terms of rankings given by the purposiveness criterion. It may, of course, be necessary to follow a similar procedure in terms of other presumptive criteria. And it must not be forgotten that the several value criteria are presumed to have an utterly equal claim on the values they sort. To give one criterion a presumptive 'lead' role is to provide the occasion for one of the most frequent objections to moral theories: what might be called the fullness objection. Consider a situation in which different criteria give the same high ranking to diametrically opposed values for an x. To give the ranking of, say, the purposive criterion more 'weight' than the aesthetic one *without having secured some reason for doing so* will raise the justifiable objection that the solution to an otherwise deadlocked situation does not do justice to the 'fullness' of human experience and value—that a large element has been slighted. To give the ranking of the aesthetic criterion the extra weight is likely to 'defy common sense.' The criteria are thus to be treated as equals until some reason can be advanced for giving one or two a special status.

Difficulties do arise in regard to additivity, of course. Values accumulate in the sense that two values of the same rank outweigh a third (and opposite) value of the same rank. But it does not appear that there is a way to add values so as to decide, for example, whether or not two values of an inferior rank outweigh one (opposite) value of a superior rank. Any assignment of numbers such that (formally) a decision can be rendered on such questions seems bound to be no more than arbitrary, due to the lack of discoverable properties in the data (values) to correspond to the additivity properties of the number system. This situation confounds the attempt to 'balance' values in many cases, and contributes, again, to the so-called looseness of moral arguments.

But the failure of the procedure to be a decision procedure in all cases should not be confused with assertions about the impossibility or non-existence of procedures which are sometimes effective. The procedure just outlined will be effective whenever a rank ordering will suffice as a balancing device—i.e., whenever the decision between opposite values for the same conduct option or between competing conduct options involves only questions of rank ordering and accumulations within a rank.

Most situations are complex, however, involving not only values of different ranks and for different conduct options, but also values that are given priorities by two or perhaps all three presumptive value criteria. Questions of commensurability between the criteria thus arise: does a value of the highest aesthetic priority 'equal' one of the highest purposive priority? Do two values of one 'second rank' accumulate to 'outweigh' one of another 'second rank'? As noted above, here again the presumption device is useful: the ranks are taken to be both commensurable and equal unless some reason can be secured for concluding otherwise. In the case that a reason can be offered for concluding simply that the ranks are unequal, of course, they will be assumed commensurable and the inequality will cause no difficulty. (For then one will simply have a situation in which a rank of one criterion is superior to a rank of another.)

In case it can be shown that the criteria are not commensurable, all hope of deciding cases involving opposite values for the same conduct option under different criteria is lost, as is hope for deciding cases involving opposite values for different conduct options under different criteria. But even so, cases involving only pro or con values for a given conduct option are decidable. That is, suppose of the given conduct options, one and only one of them has 'pro' values (and more of them than any other option) in the highest rank for each presumptive criterion in use. There is no difficulty then in deciding which of the options is the best. While it may seem at first that such fortunate cases are likely to be few and far between, a bit of reflection suggests that some of the cases considered most crucial in the testing of a moral theory (gratuitous homicide, cruelty, etc.) are precisely of this sort. In any case, the procedural point is here that if a reason can be secured for concluding that the ranks of the several criteria are incommensurable, it is still possible to render a decision, in some cases, on the matter of what conduct option is the right one. Of course, as long as the presumption in favor of commensurability remains, decisions can be rendered in many more cases.

3 SOME REMINDERS AND ELABORATIONS

It bears repeated emphasis that the ranking operation performed by the presumptive value criteria are *not* to be construed as valuations. They are dispositional sortings which need not be given valuational status at all—indeed, cannot be given such status unless one is prepared to talk about 'subconscious' or 'operational' valuations which have no hint of being conscious acts. It thus follows that the procedures outlined here *do not* propose putting the results of all first order valuations into the maw of a second order set of calculative valuations—e.g., *x* value is 'good for' purposiveness, etc. No doubt for some purposes, the operation of the value criteria can be described by analogy to calculative valuation, but the description is only an analogy, and so it is not the case that these procedures entail that 'all valuations are at bottom calculative' or some such reductivist thesis. Conceptualization enters the picture as an adjunct to the operation of the criteria, with the presumptive justification offered for their continued use, the retrospective and prospective use of the criteria, and the propositional formulation of them. This point needs emphasis because when one describes the criteria in operation so as to include them in reasoned justifications, one can easily get the idea that there is some calculation of consequences going on in their operations.

Further, of course, the complexity of many value situations has the consequence of leaving one 'not knowing which way to turn'—in this case, getting no help at all from the presumptive value criteria. The values simply do not 'fall into' a rank order. What happens in such cases is instructive as to the way the value criteria operate: we redescribe the situation, we reformulate the valuations, we redefine the value criteria—all efforts directed at squaring things off, getting things to 'fall into place.' This is not (or need not be) described as valuing. What it amounts to, operationally, is a trial and error method of 'making the situation fit,' seeing it in a way that makes it obvious how the criteria rank it. And of course this opens the gates for disputation again—for an argument over the adequacy of various formulations (You are only giving an *ad hoc* description designed to *force* the facts to fit. . .). For there is nothing in these procedures which dictates that all facts *will* fit.

Further still, it should be noted that one may in some cases just have to act in order to get a ranking by the criteria. That is, the criteria are not always adequate prospectively, and thus a certain amount of learning by doing is inevitable. The 'high values' of our culture come in large part from the context of the widest and deepest human experience, and, as is often said, one without the sensitivity or openness to this experience can hardly understand why the 'high values' are ranked as high. This may be seen as due to the prospective inadequacy of the criteria.

Finally, one often acts purposively, or as a person, or from inclination *without having valued* the course of conduct taken. That is, the features of life here treated as value criteria also operate as motives for action independently of what has here been called valuation. This factor is another way in which their operation may seem to some covertly valuative, but a motive need not be construed as a valuation or even as 'containing' one.[2]

4 SOME SOURCES OF UNEASINESS

A few loose ends remain to be picked up before the discussion moves on to other topics. For one thing, the 'cumulativeness' asserted for values may be questioned. Even granting that the procedures so far introduced do not make the mistake of asserting additivity for rank ordered values, it may be argued that the uses made of accumulations of values are equally as mistaken. After all, it does seem silly to say (as the procedures suggest one might) that because one object is valued more often than another it has more value than another—that is, that one value for x, plus another value for x, results in x having more value than if only one value had been secured for it.

Part of this difficulty, this seeming silliness, is verbal, but a hint of the old subjective/objective issue lies behind the verbal quirk. If one insists on discussing only the values of objects, and further is led to believe that the value of an object is necessarily a 'unity,' the accumulation of values will make no sense to him. 'More value' means 'greater value,' 'higher value,' in his lexicon, and he thus wants to know: higher value of what? Is the happiness of two people greater happiness than the happiness of one? Whose happiness is this 'greater happiness'?

The answer to be made from the present account has been suggested earlier. It is simply that one cannot rule out values *for* objects in deciding what is best to do, and the inclusion of 'values for' means a plurality of values for most, if not every object. Once this point is clear it is relatively easy to deal with the verbal quirk, which is that the 'silliness' of saying that more valuations equal more value comes partly from the use of the singular noun 'value.' One apple plus another apple does not mean that a man has more *apple* than if he just has one; it means he has more *apples* than if he just has one. Similarly, one value for x plus another value means x has more *values*. And it is taken for granted that a conduct option can be 'best' just as surely by having the *most values* as by having the *greatest* (highest) *value*. So the accumulation issue can be laid to rest.

Some dissatisfaction and unease may also be expected concerning the precariousness of the grounding procedures. If it can be shown that there are good reasons for abandoning the presumptive value criteria, then one is surely back where he started with respect to the grounding problem. No mitigation of such unease seems possible, though in the absence of specific suspicions about the criteria one should not be overly concerned. The precariousness here is no more fundamental, intellectually, than that of fundamental theories in science. It seems more crucial, of course, because a mistake in conduct is apt to have more than purely intellectual consequences. But perhaps some solace can be taken from the fact that what we require of a moral man is not that he make no mistakes, but that he not undertake to do what he knows is wrong, and that he make every effort to discover what is right. Conduct is often perilous morally, and one must take his chances: that he will actually bring off what (and only what) he sets out to do; and that what he has decided to do is of the value he thinks it to be.

This last phrase brings to mind again the pervasive 'looseness' and 'openness' of the procedures so far outlined. They do not march the user inexorably to necessary conclusions, but rather shamble towards more or less secure judgments. They are not decision procedures, then, but procedures which occasionally yield decisions. The fact that they can yield decisions, however, is of considerable significance. It means that the grounding problem is not in principle unsolvable. And if the solution in principle proves to be less than universally effective for practice, this should not be at all surprising. The suspicion that some moral problems are simply intractable is not new, and it is by now a platitude that reasoning has severe and rather well-defined limits.[3]

Finally, one may wonder whether the fact that the features of life used as presumptive criteria rarely occur in our experience standing alone, as it were, militates against using them in the way recommended. That is, they are separately describable, but are they separately operable? If they are not, can they still be used as presumptive criteria?

The answer to the latter question is yes. The criteria then operate as a single criterion. The answer to the former question is that, certainly in situations of conflict, the three features are separately operable. When there is no conflict, there is no need to treat them as separate in any case.

NOTES

1 The distinction noted here between ordinal scales (those possessing only the rank ordering properties of the transitive, asymmetrical relations $>$, $<$, and the transitive symmetrical relation $=$, and the intransitive symmetrical relation $\neq$) and the interval and ratio scales which have additional properties (notably additivity) is the standard one made in the literature on the theory of measurement. See: S. S. Stevens, 'On the Theory of Scales of Measurement' in A. Danto and S. Morganbesser, eds, *Philosophy of Science* (Meridian Books, 1960), pp. 141–9; M. Cohen and E. Nagel, *An Introduction to Logic and Scientific Method* (N.Y.: Harcourt, Brace & World, 1934), pp. 289–301. Stevens's article gives a particularly succinct summary of the statistical operations possible with the various scales. The reader will also want to consult Rescher, *Introduction to Value Theory* (Englewood Cliffs: Prentice-Hall, 1969)—particularly for Chapters VI and VII and the extensive bibliography. Rescher argues that the ordinal versus cardinal scale differences have been overemphasized. If this is true, and if a cardinal number scale can be constructed in these matters, the power of these procedures would be increased enormously. Other helpful articles: G. N. Belknap, 'The Commensurability of Values,' *Journal of Philosophy*, 30 (1933), pp. 458–64; P. M. Brown, 'Distribution and Values,' *Journal of Philosophy*, 66 (1969), pp. 197–210; John C. Hall, 'Quantity of Pleasure,' *Proceedings of the Aristotelian Society*, 67 (1966–7), pp. 35–52; and Joseph Mayer, 'Comparative Value,' *Philosophical Review*, 45 (1936), pp. 473–96.

2 For a similar 'non-valuational' account of the operation of value criteria (though no reference is made to their presumptive nature), see Stephen C. Pepper, *The Sources of Value* (Berkeley: University of California Press, 1958), Chapter 13, section 4.

3 Articles stressing the looseness of moral arguments: J. Cohen, 'Three Valued Ethics,' *Philosophy*, 26 (1951), pp. 208–27; J. N. Findlay, 'The Methodology of Normative Ethics,' *Journal of Philosophy*, 58 (1961), pp. 757–64; G. C. Kerner, 'Approvals, Reasons, and Moral Argument,' *Mind*, 71 (1962), pp. 474–86; D. J. O'Connor, 'Validity and Standards,' *Proceedings of the Aristotelian Society*, 57 (1956–7), pp. 207–28; Jerome Stolnitz, 'Notes on Ethical Indeterminacy,' *Journal of Philosophy*, 55 (1958), 353–67. R. B. Braithwaite, in 'Common Action Towards Different Moral Ends,' *Proceedings of the Aristotelian Society*, 53 (1952–3), pp. 29–46, points out, however, that agreement on what to do is not necessarily blocked by disagreements about principle; and H. N. Wieman (in 'A Criticism of Co-ordination as a Criterion of Moral Values,' *Journal of Philosophy*, 14 (1917), pp. 533–42) makes the interesting point that total agreement is not an unmixed blessing, anyway.

IX

MATTERS OF MORAL CONCERN

The distinctions between moral and non-moral judgments, moral and non-moral values, moral and non-moral practices are, on the face of it, quite relevant to the arguments of this book. The aim here is, after all, to find procedures for grounding moral judgments, not necessarily other kinds. And yet no attempt has been made to say what a moral judgment is, as distinct from other kinds. Indeed, in terms of the painstaking studies done on types of value expressions, distinctions between rules and principles, and the careful way in which the 'logic of moral language' has been approached, the foregoing arguments must seem somewhat cavalier. Care on these matters here has been largely behind the scenes. It might be wise to make some of it explicit before going on.

1 PRELIMINARIES

It is, for one thing, understood throughout that what is justifiable, justified, to be submitted to these procedures, to be grounded, to be secured. . . are assertions (statements, propositions, judgments). The aim here is to establish procedures for reasoning to conclusions which answer at least some of the questions raised by moral problems, and such conclusions, as well as the premises which secure them, are to be sentences capable of being true or false. So even though various shorthand locutions have been used (e.g., securing a valuation), it is to be understood that they are only shorthand. Further, 'judgment,' 'assertion,' 'statement,' and 'proposition' are here used interchangeably. While there are obviously contexts whose subtleties call for distinguishing, say propositions from statements, the need does not make itself felt here. Similarly, nothing crucial in these arguments hangs on a distinction between rules and principles, maxims and laws, judgments and principles.

It has been noted before that expressions of three of the types of valuation can be straightforward assertions. Expressions of affective valuations are not, and are not secured by, reasoned argument; but each value expression is easily correlated with an assertion *about* the affective valuation, and this assertion *is* open to reasoned argument, and can enter as a premise into further arguments. Expressions of the *e*valuation of valuations—that is, the balancing of goods and bads to determine the best course of conduct—are also straightforwardly assertions. So the existence of the proper sort of 'objects' for the use of reasoning procedures (i.e., sentences that can be true or false) is assured at each stage.

2 THE MORAL AND THE NON-MORAL

What remains then is a discussion of the difference between the moral and the non-moral as it relates to the arguments of this book. It has so far been assumed that while moral values may well be a special sort of value, they are not ones which need special grounding procedures. But if it were to turn out that there is a radical difference of type here—that moral judgments have a 'logic' of their own—much of the work so far done on general theory of value would be suspect: one would wonder how much, if any, relevance it has to moral values and moral judgments. So it will be necessary to inquire into the nature of the difference between the moral and the non-moral, at least to such an extent that one can be sure the difference is not a crucial one for matters of procedure. Hopefully, the inquiry will not require the production of a full dress set of criteria for distinguishing moral from non-moral judgments, values, or whatever. For as is the case with many distinctions that give people little trouble in the using, this one is very difficult to formulate—that is, to express in the form of a definition or a set of criteria.

Philosophers have approached the issue from several angles—or perhaps one should say, have dealt with several related issues. They have tried, for example, to distinguish moral from non-moral principles, moral from non-moral conduct, and moral from other types of responsibility, e.g., legal responsibility.[1] As is to be expected, different questions elicit somewhat different answers. A discussion of moral principles or judgments must consider the question of whether or not moral and non-moral judgments have formal, linguistic differences. Discussions of moral (as opposed to non-moral) conduct emphasize differences of motive and differences of response to various sorts of conduct. 'Judgment accounts' are likely to deal with matters of universalizability, the relevance of certain kinds of excuses when the judgments ascribe responsibility, and the primacy or importance of the judgments. 'Conduct accounts' are likely to discuss the responses of conscience, guilt, righteousness, and 'the good will' in making the distinction. But it will not be surprising if some significant features of all these analyses—as well as the one to follow—are found to be similar. In particular, it seems that most analyses, from whatever questions they begin, find that moral values, judgments, conduct, or rules are distinguished from other types by being somehow more important—e.g., 'over-riding' and more consequential in human affairs. The question asked here is how one is to distinguish moral valuations (and hence moral values) from non-moral valuations. The answer to this question will also turn out to make a good deal of the special 'importance' of the moral as opposed to non-moral.

3 MORAL JUDGMENTS ARE ONES REQUIRING A GROUND

An operationally effective criterion for determining whether a 'context' is a moral one or not, whether a judgment is a moral one or not, or whether a value is a moral one or not, is very nearly provided by simply asking about the insistence with which the grounding question plagues the reflective people involved. To explain: there are contexts (satire, sick humor, fantasy) in which the grisliest 'events' and the most dogmatic value pronouncements do not raise any sort of challenge from reflective people. A calculative valuation of a hammer in the simple context of nailing together a bookcase, while it may raise a protest, hardly calls forth a demand for a grounding, at least as that notion has been developed in the preceding pages. Contexts that *do* raise the impulse—in thoughtful people—to ask for more than 'some reason' to believe a value judgment (i.e., which raise the question of grounding those judgments) are ones in which we would say we were morally concerned, in which moral issues were involved, in which moral values were at stake. And value judgments made in such contexts—when we insist on grounding them—are moral, as opposed to non-moral, judgments—whether they indict someone for wrongdoing, or state a principle, or express a valuation, or an evaluation.

In fact, it looks very much as though the contexts which raise the grounding question for thoughtful people are *precisely*, exhaustively, the ones we want to call moral. So that determining that the grounding question is relevant amounts, operationally at least, to determining that the context is one in which people are morally concerned, and that the value judgments involved are moral ones. This may not *explain* why we call these contexts, values, and judgments moral ones. Indeed, when one asks people why a value is a moral one it is unlikely that they would even consider such a reason. The criterion is only put forward here as an operationally effective one. But even if it is only that, and if it is correct, it assures one that a study of grounding procedures will not suffer from ignorance of the why: all the judgments one is concerned to deal with for purposes of grounding will be moral ones. And to discover enough about the 'logic' of judgments to get them grounded will be in fact to discover at least part of the logic of moral judgments—by definition, enough of a part for present purposes.

Yet some suggestions are in order about why it might be that this criterion is operationally effective. For example, when is it that reflective people seem to insist on the relevance of the grounding question? And does an examination of the occasions on which they are thus insistent give any clue as to why we are willing (what we apparently mean to indicate when we are willing) to call the situation a moral one?

Suppose one takes the notion of valuation and asks what features are required for our willingness to say that a given valuation is a 'matter of moral concern,' to say that what is at stake is a moral value. It appears that the features (of the context, if you like, or the valuations' ramifications) which lead us to describe it as a moral matter—resulting in a moral value's being of moral concern, being such as to call forth the moral responses (e.g., anxiety, guilt, remorse, shame, and their opposites)—are all matters of degree, and may all be described as involving ways in which valuations can assume 'importance' in our affairs.

To be specific, it appears that a valuation may be described as a matter of moral concern to the degree that it (1) is involved in a need to get something done; *and* (2) is involved with significant life-consequences; *and* (3) is involved with the practice of holding people responsible (liable for sanction); *and* (4) is involved with the issue of a man's character.[2] To develop these suggestions fully would take more space than sticking to the subject at hand will license, but briefly:

Sometimes when a valuation is made, it is appropriate (i) to interpret it as, in part, a call for something to be done; or (ii) to ask whether anything can or ought to be done 'about' the matter. And at other times neither is appropriate. Consider a valuation in aesthetics. One says, 'That painting is wonderful.' Someone else says, 'I don't understand what you want me to do about it.' Or again, 'Ought anything to be done about it?' Even admitting the vagueness of 'doing something about it,' it still is the case that if neither of these questions can be judged appropriate to what was said (i.e., if neither (i) nor (ii) is at all 'called for') then it is surely very odd to call the valuation a moral one. The appropriateness of the question does not guarantee the use of the description 'moral'—i.e., is not a sufficient condition for it—but it does seem to be necessary for it. And of course a valuation's 'involvement' with a need for action can be a matter of degree, just as the need itself varies in degree.

As for item (2), it surely appears odd to describe, as moral, a valuation that has nothing to do with significant life-consequences —either directly or indirectly. The 'value-choice' between Wheaties and Corn Flakes can only be a moral matter to the degree that it can be shown to make some difference—e.g., to violate one's responsibilities if the 'wrong' choice is made.

Similarly for items (3) and (4): some valuations are involved with these matters and others are not. When they are, and to the degree that they are, we do not resist calling them moral. When they are not, and to the degree they are not, we do resist calling them moral.

The applicability of the description 'moral' is not, of course, a rigid matter. Not all these 'involvements' must be met to a specified degree to move one to use the description. But the presence of all does seem to be required for the legitimate use of the description, and further, each of these items seems to be getting at a way of saying that the valuation so involved is 'important.' That is, each of these items expresses a notion we often have in mind (mean) by saying that a thing is important.

Is to say, then, that a valuation is a matter of moral concern merely to say that it is an important concern in these ways? It is doubtful that the matter is quite so straightforward in its entirety. Even so, this partial analysis suffices to give at least provisional assurance that the inquiry into grounding procedures may proceed without a 'final' answer on the matter. For if, in effect, all the value judgments we are concerned to ground are the moral ones, to have exhibited value grounding procedures will be to have exhibited procedures for moral judgments, and the argument here will not have suffered in validity by proceeding without a full-scale analysis.

NOTES

1 Literature on the universalizable criterion for the distinction includes: R. M. Hare, page 31 of *The Language of Morals* (Oxford: Clarendon Press, 1952), as well as many passages in *Freedom and Reason* (Oxford: Clarendon Press, 1963); Jonathan Bennett, 'Moral Argument,' *Mind*, 69 (1960), pp. 544–9; and A. C. MacIntyre, 'What Morality is Not,' *Philosophy*, 30 (1955), pp. 256–65. On the 'feelings' approach, see C. L. Stevenson, 'The Emotive Concept of Ethics and Its Cognitive Implications,' *Philosophical Review*, 59 (1950), pp. 291–304. For other interesting articles, see T. L. S. Sprigge, 'Definition of a Moral Judgment,' *Philosophy*, 39 (1964), pp. 301–22; S. S. Brown, 'Right Acts and Moral Actions,' *Journal of Philosophy*, 42 (1945), pp. 505–15; Herbert Morris, 'Punishment for Thoughts,' in R. Summers, ed., *Essays in Legal Philosophy* (Oxford: Basil Blackwell 1968), pp. 95–120. For a discussion of the distinction between moral and legal responsibility, see H. L. A. Hart, *The Concept of Law* (Oxford: Clarendon Press, 1961); and Lévy-Bruhl, *L'Idée de Responsabilité* (Paris: Librairie Hachette, 1844).

2 When one considers the distinction between moral and non-moral character traits, some adjustments will have to be made in this analysis. Specifically, item (4) gets dropped and items (1), (2), and (3) are reformulated: 'A character

trait may be described as a moral (as opposed to non-moral) one to the degree that it (1) involves dispositions toward some sorts of conduct and away from others; *and* (2) the conduct options involved have significant life consequences; *and* (3) those conduct options are involved with the practice of holding people responsible.' It may be argued that all character traits involve (1). But not all involve (2) or (3). Further, it should be noted that nothing in this analysis prohibits *retrospective* moral judgments. Only the verb tenses change.

X

GROUNDING DEONTIC JUDGMENTS

It can hardly be doubted that the concept of obligation is central to morality—at least in our culture, and probably in all highly developed social orders. The justification of statements of obligation (hereafter: obligation-statements) must thus be a primary concern of any general inquiry into justification procedure.

Two routes for grounding justifications of obligation-statements will be outlined here: one by way of the axiological scheme already presented; the other directly from the notion of an obligation-making institution to some presumptive deontic criteria. In the former procedure, the problem to be examined is the traditional one for axiological schemes—the one of getting from is to ought. For if a linkage can be provided between value judgments and obligation-statements, then the grounding procedure for value judgments serves, indirectly, for obligation-statements as well. In the latter procedure, the problem to be examined will be that faced by all deontological schemes—that of justifying at least some obligation-statements independently of the balancing of values process.

It should be emphasized that these two routes are not put forward as *alternative* procedures. They are put forward as *complementary* procedures—part of the effort, as mentioned in Chapter II, of co-ordinating axiological, deontological, and agent morality approaches to moral theory.

1 FROM IS TO OUGHT

The procedures so far outlined take one from the valuation of conduct options through their evaluation—i.e., through the balancing of goods and bads to determine which conduct options are right to do. So far 'right' has been defined as 'the best of the available options.' This is unobjectionable enough when one is concerned with developing an axiological scheme. But when one looks at the conflict between axiologists and deontologists, he soon sees that such a straightforward definition of 'right' will not do—not, at least, for any theory which hopes to co-ordinate the two approaches.

To see this clearly, consider: 'One ought to do what is right' is held to be analytically true in the sense that the notion of a thing's being right is said to be 'contained' in the notion of its being what one ought to do. That is to say, whatever else can be said about the things I ought to do, it will always be true to say of them—simply from the fact that they are what I ought to do—that they are 'right' to do. But the deontologist is wary of any claim, here, that the 'right' thing to do—when it is thus derived from the ought—must necessarily indicate that the balance of values in some axiological scheme favors the act in question. He is prepared to say that if an act is one's obligation, then it is right to do regardless of any value-balancing operations. I suggest that the axiological scheme proposed in this book blunts *most* of the reluctance one has to define 'right' as the best of the available options, but still, it is important to note the ambiguity here, for it may not be altogether eliminable.

In any case, both parties to the dispute will admit that arguments of the following form are trivially valid:

> One ought to do x (where x is a course of action).
> Therefore, x is 'right' for one to do.

The question for the axiologist, of course, is this: will the inference ever work the other way—from the 'is' to the 'ought'? Is it ever possible to say, from the fact that x is right to do (as an axiologist defines 'right'), that 'one ought to do x'? Suppose that:

> X is in some sense good (i.e., has a value as good).
> X is in no countervailing sense bad.

Therefore, x is in that sense right to do (i.e., doing x is, as far as one has gone, more good than bad).

This inference works given the previous definition of 'right to do'; the premises express the only considerations relevant to deciding the truth or falsity of the conclusion. But does the next one work?

X is right for one to do.
There is no better option.
Therefore, one ought to do x.

Or will it work to say:

X is the right thing for one to do.
Therefore, one ought to do x.

The reluctance to let such inferences pass unchallenged comes from quite legitimate concern over some ambiguities in the use of 'ought' and 'right,' and some obvious situations which seemingly excuse one from 'having to do' what is right. The latter concern will be dealt with by the inclusion of a *ceteris paribus* clause in all these arguments, together with a discussion of excusing conditions. The place to begin, however, is with the ambiguities.

A 'NON-OBLIGATION-SENSE' OF 'OUGHT'[1]

'I ought to do x' can express merely that the balance of values for x is, in fact, 'more good than bad,' or that x is the best thing to do. In other words, 'ought' can be and is used without any trace of the notion of obligation, and in a way that can only be paraphrased by reference to what is right to do, as 'right' is here defined. 'I ought to do x,' used in this way, is merely another way of indicating one's conclusion about what is right to do. This is not an *ad hoc* sense of 'ought' invented purely for present purposes, but a genuine sense which has a significant usefulness in everyday as well as technical discourse.[2] Its usefulness is to call attention to the way in which doing what is right may be 'urged' on one by factors other than the estimate of the value of the conduct.

To explain: deliberations as well as choices or decisions about what course of action to take are purposive events. It is a part of the description of purposive activity to say that it 'tries' to achieve satisfaction. It is conative activity—striving—and has momentum. In the case of trying for a conclusion about what is right to do, this momentum of the trying (though 'logically' unrelated to the subject matter under consideration) may get imported into the conclusion as a sense of urgency about going on with it—about putting the conclusion into a decision and acting on it. It imparts, in other words, an emphatic force to '*x* is right' which one can express by changing the words to '*x* ought to be done!' The emphasis can get itself expressed in other ways, of course. But 'ought' is peculiarly attractive for this purpose, due perhaps to its being the old past tense of the verb 'to owe,' and its use in obligation statements (which are, of course, an even more significant way of going beyond the assertion that a thing is right merely by virtue of a balancing of values). The point is merely that it can and does get expressed by the change from 'right' to 'ought.' To say, then, that one ought to do what is right may only be to say that things are already in motion and to resist doing what is right would be to back things up—to call a halt—to ignore the urging in the existing momentum. Correspondingly, to say that one ought *not* to do what is right can mean that there is reason to resist the existing momentum.

If no reasons can be found for resisting the momentum, then the importation of the emphasis in the switch from 'right' to 'ought' is philosophically innocent at worst. It should be emphasized, of course, that it is the absence of reasons to the contrary which makes the 'urging' philosophically innocent. To one who objects: But what is the reasoned connection between the mere evaluation (*x* is right) and the emphatic element in the ought statement? What justifies making the transition? the reply must be that no justification is necessary if there are no reasons to the contrary. If there are no reasons against, the possibilities are two: (1) there are also no reasons in favor, and thus the transition to the ought statement is a matter of reasoned indifference; or (2) there *are* reasons in favor, in which case, given the absence of reasons against, the move is justified. In either case, no reasoned objection can be made, and the argument proceeds.[3]

Now, what could count as a 'reason against' making the move? Obviously, an evaluative judgment against its appropriateness: a valuation of the move as bad with no countervailing goods—i.e., a judgment that the move is wrong. Such a judgment need not itself proceed to an ought statement (i.e., to the statement 'one ought *not* to do *x*') in order to block the reasonableness of the original move from right to ought. And the reason to the contrary need not be at all a contradiction of the original evaluation of the 'rightness' of *doing* *x*. It is an evaluation of the rightness of *urging* *x*.

When there are no reasons which can be secured against importing the emphasis into the assertion that *x* is right to do, the emphasis will be called innocuous. Inferences of the following forms, then, can go through:

(a) *X* is right for one to do.
There is no better option.
Therefore, one ought to do *x*.

(b) *X* is right for one to do.
X is the only course of action which (in these circumstances) can be justified as more good than bad.

(c) X is the right thing for one to do.
Therefore, one ought to do *x*.

These inference forms are acceptable as long as 'one ought to do *x*' expresses no more than the 'rightness' of *x* and the innocuous emphasis, for in that case the premises express the only considerations relevant to deciding the truth of the 'ought statement'; the rightness issue has been decided, and there are no reasons against importing the emphasis.

The curious feature of all this is that it occasionally happens that when one has undertaken to find out what to do and has reached a deadlock in that all the available conduct options balance out as equally good (or bad), the momentum discussed above can get imported to one and only one of the options. The result is that unless some reason to the contrary can be given, the one which gets the momentum may be said to be the one which ought to be done. And so deadlocks sometimes can be broken with this form of the move from 'right' to 'ought.'

THE 'OBLIGATION-SENSE' OF 'OUGHT'

'One ought to do *x*' can be taken to mean 'one is obligated to do *x*.' If this is the sense of the ought statement it surely is not the case that inferences of the forms (a), (b), and (c) above will work. It does not follow (for example) from the fact that '*x* is the only thing around which is on balance more good than bad' that 'one is obligated to do it.' At least it is not clear that one can be, from that fact alone, under an obligation. One may recall arguments to the effect that he is obligated to maximize good—both for himself and others. When this premise is added to '*x* is right' then it does indeed follow that one is obligated to do it. But this is obviously not a case of inferring an obligation to do *x* from the fact that *x* is right. It is rather a case of inferring an obligation from the fact that *x* is right *and* from the principle that one is obligated to maximize good. So, again, at least one can say it is not clear that an obligation to do *x* follows from the fact that *x* is right.

What is clear is a way in which obligations may be inferred from 'institutional fact.' Professor Searle's admirable article[4] outlines a procedure for such inferences very carefully. The procedure suggested by Searle, in outline, is first to mark a distinction between two kinds of fact—institutional and brute fact. Next to note that the institutions involved in institutional facts may be expressed as systems of rules. Then to specify how a person puts himself, or can get placed, 'under' an obligation-making institution, and finally to show how all of this may entail the judgment that a person ought (is obligated) to do *x*.

To explain: some statements of 'fact' cease to be capable of being true or false when the corresponding human institution (set of practices, agreements, expectations) disappears. Searle's example relates to money. 'She has five dollars' is neither true nor false unless a dollar money system is presupposed. The system need not be currently working to make the statement false; but it certainly needs to be around to make the statement true. And the truth or falsity of the statement always depends on human beings' having once dreamt up the 'institution' of dollar money. Other statements of fact which, following Anscombe, might be called statements of brute fact do not depend on human institutions in this way at all.

Both types of 'fact' are stated with the 'usual copulations of propositions'—the verb 'to be'—but at least some statements of institutional fact, together with other circumstances, license ought statements in a way that statements of brute fact (apparently) do not. These institutions are what might be called obligation-making institutions. The practice of promising is the usual example, since other obligation-making institutions can, at least in part, be considered special cases of this: e.g., legal contracts. It is a commonplace that such practices can be given a descriptive definition in terms of a set of rules: so-called constitutive and regulative rules. The constitutive rules define the conditions under which promising may be said to have occurred, and, implicitly, the conditions which prevent an attempt at promising from 'coming off.' The regulative rules define what moves may and may not be made within the frame defined by the constitutive rules. Clearly, such rules may quite unobjectionably be stated in terms of what, given the institution's 'constitution,' it is right to do, and one may thus easily move to statements of what one ought to do in the sense described above under (1).

The notion of an obligation, and consequently of the obligation sense of some ought statements, comes, it would seem, from the notion of bounding a man's choices by these special institutions. Once such a decision has been reached, any ought statements which may result take on an especially compelling force: they restrict one's range of conduct options by ruling out any available ones which do not 'fit' with the institution—either by following from it or by being consistent with it. The usual metaphors for obligating situations come quite naturally from this: being 'bound' to the obligatory course of action, legally binding contract, etc.[5]

The more one concentrates on the way in which his options are bounded by the institution, the more emphatic 'force' discussed above focuses on what, in terms of the institution, it is right to do. The 'sense of obligation' thus is plausibly explicated as the emphasis plus the awareness of the boundaries that 'bind' one to concluding that the best thing in terms of the institution is *the* right thing to do.[6] (Without the decision to bound one's options by the institution, of course, a conclusion that x was best in terms of the institution, or good as an 'institutional act,' would not necessarily be a conclusion that x was *the* right thing to do.) Obligations may conflict in the sense that two or more institutions by which one's options are bounded entail conflicting ought statements.

AN OBJECTION CONSIDERED

Recently, Hubert Schwyzer[7] has constructed an argument against the contention that practices can be adequately characterized by rules, and since he mentions Searle's 'How to Derive "Ought" from "Is"' as an article his arguments have force against, perhaps it should be mentioned why they do not destroy the use I wish to make of Searle's thesis. Schwyzer argues that 'the nature of a given practice is defined not by its rules but by its "grammar," by what things it is relevant to say and do with regard to it . . . Rules (constitutive rules) do not themselves specify how the behavior in accordance with those rules is to be regarded; that is something that the very setting up of the rules must presuppose.' He urges this point by constructing an example to show how people might behave in accordance with the rules of chess but not be 'playing chess' at all. He then concludes that 'the game of chess is not adequately characterized as a "system of rules." Playing chess does not *consist* in acting in accordance with the rules. The rules do not explicate the concept of playing chess; they do not establish what it is to play chess.'

It would be a mistake to think that this argument has force against the procedures outlined here. What Schwyzer's example shows, strictly, is that the activity of playing chess is not adequately (i.e., exhaustively) described by what we ordinarily agree to be 'the' rules of chess. He has not shown that 'the' rules of chess (at least those we call constitutive rules) are not necessary parts of an attempt to define the nature of the activity called 'playing chess.' Indeed, one assumes he would at least agree that 'the' rules of chess *might reasonably be included* in a characterization of chessplaying, since his example makes use of them.

Schwyzer seems to assume that the arguments by Searle depend on holding that an activity or practice can be adequately (i.e., exhaustively) defined, described, or characterized by listing what we would ordinarily agree to be *the* rules of the activity (e.g., in the case of chess, *the* rules of chess). This is manifestly not the case, at least in so far as I understand the arguments. Regardless of how Searle's points are (literally) expressed, they depend at most upon the contention that an adequate characterization or definition or description of a practice (institution) will *include* 'the rules' (constitutive rules) of the practice (if it is the sort of practice that has explicit constitutive rules), and that a good deal of what Mr. Schwyzer calls the 'grammar' of the practice *can* be described in terms of a set of rules. Searle need not hold either that *the* rules of a game exhaustively define its 'nature' nor even that there *is* a set of rules which does this.

GROUNDING OBLIGATION STATEMENTS

So it should be clear that the inference from 'is' to the (obligation) 'ought' causes no formal problem as long as some of the 'is' statements can be shown to be statements of institutional fact—for these, when explicated in an appropriate way, involve some 'ought' statements themselves. What *is* a problem, and one explicitly raised by Searle's article, is this: supposing I can infer statements of the form 'I ought to do *x*' (meaning 'I am obligated to do *x*') from statements of institutional fact, why am I obligated to abide by those rules? Why 'ought I' (am I obligated) to define the institution in the way it is presently defined? Why can I not define it in such a way that no obligation would fall on me from it? To answer that there is another institution from which an obligation follows to define this one in a certain way only pushes the question back a notch. There is an indefinite regress possible here. The problem is not really solved by Searle's footnote[8] in which he says that one could not 'throw *all* institutions overboard' and 'still engage in those forms of behavior we consider characteristically human.' For the issue is not an 'all or nothing' one—at least not in its most perplexing form. Nihilism may perhaps be dealt with in this way, but surely the crucial questions surround *which* of the institutions (if any) we can (or ought) to change, how they ought to be changed, and whether *any* (rather than *all*) can be 'thrown overboard.'

Thus the way in which one can ground the justification of the assertion that he 'ought' to define an institution in a certain way cannot be by showing that he is 'obligated' to do so. But he may be able to decide that this is what he ought to do in the first sense of 'ought.' He may be able to decide that, on balance, doing *x* is the thing which is (the best of those or the only one which is) more good than bad, and that he has no reasons for resisting the momentum toward doing what is right (no reason for resisting the importation of the emphasis—the switch from 'right' to 'ought'). Thus he may be able to proceed as follows:

(i) *X* (the act of bounding a man's options by the rules of a certain institution) is good.
X is in no countervailing sense bad.
Therefore, *x* is right to do.

(ii) *X* is right for one to do.
There are no reasons against importing the emphasis.
Therefore, one ought to do *x* (meaning only sense (1) of 'ought').

(iii) The institution to which a man's options are bounded by *x*, when descriptively defined by a set of constitutive and regulative rules, entails the consequence that given conditions A, B, and C, the right thing for him to do is *y*, and barring reasons against importing the emphasis he ought (in sense (1) of 'ought') to do *y*.

Argument (iii) is an abbreviated version of Searle's more careful form for deriving an ought from a statement of institutional fact, supposing that one's options have been bounded by the institution. Arguments (i) and (ii) consider what Searle calls 'external' questions (or, in the footnote mentioned above, 'tinkering' with institutions). I suggest that arguments (i) and (ii) exhibit a procedure for dealing with the question of which institutions I 'ought' to assent to or how they 'ought' to be defined, or why we 'ought' to have any at all. And that all the arguments, taken together, constitute a procedure for justifying what are called obligations—i.e., the being bounded by an institution which entails ought statements of the (1) sense of 'ought.'

FINAL REMARKS ON IS TO OUGHT

It will be noticed that the notion of obligation used here in connection with the second sense of 'ought' is very much broader than that of ordinary usage and some philosophical analyses.[9] 'Obligation' has been used as a blanket term to cover the notions of duty, the more ordinary notion of obligation (which emphasizes a nearly, if not always exactly, contractual relation), and what will in later sections be called deontic responsibility. This conflation is made merely for convenience, for in terms of this aspect of the procedure question, the three need not be separated.

Further, it should not be concluded that the analysis presented here is just one more version of 'ideal' utilitarianism. True, the concepts of right and wrong, duty and obligation are treated as derivatives of 'good' and 'bad'—and so this analysis must be called axiological as opposed to deontological. But it would be a mistake to call the analysis utilitarian. An examination of the deontologists' objections to treating 'right' and 'duty' as functions of 'good' and 'bad' (particularly utilitarian goods and bads) will show that no such objections are possible here.

Deontologists say: 'Obviously good and bad are closely tied to the concept of duty, but even in some cases where no good consequences are demonstrable (indeed, even in cases where all the known effects are bad) we still would want to say that a man might have a duty. . .' Using the analysis of preceding chapters, an account can be given of why we might want to agree. Namely: not all valuations concern consequences, and so it might be possible to hold (value) the having of a duty as good and right even though the consequences were indifferent or positively adverse: e.g., summary, good-of-its-kind, or affective valuations in favor of a duty which utility rules out. This makes clear at once 'why we want to say' the duty exists and gives an account of the way in which the valuations are 'closely tied' to the notion of duty. Further, if the deontologist were to object that there are cases in which *all* the valuations are adverse, but people would *still* want to say that the duty exists, the explanation to be made on this analysis would be that the existence of the role or institution which entails the duty might conceivably be justifiable even though all the valuations of a particular act entailed by that institution are negative. In that case the deontologists may be understood to be arguing that the act (duty) entailed still 'must be done,' at least in part because it can be justified indirectly as a consequence of an institution which it is right to bound men's options by. When one adds to this the procedure for going directly to presumptive deontic criteria in justifications of obligation-making institutions (to be developed below), it seems to me that the demands of deontologists for justifications not exclusively controlled by an axiological scheme have been met.

The point at issue in further argument will be this: must an (otherwise justifiable) institution be redefined if it entails (otherwise) *un*justifiable duties? The answer must be: surely, if possible. But it is just the poignancy of institution-building that occasionally we are not able to construct any institution for a given (highly justifiable) purpose which yet entails no such unfortunate results.

When such a situation arises we perhaps have an irreducible disagreement between those committed to duty at all costs and those willing to blink at a breach of duty—a disagreement of disposition here that no reasoned approach can assail. On this analysis, to fail to perform such a duty would be to fail to perform a 'real' but unjustifiable duty. While the deontologist might not be happy with the implied encouragement to violating such a duty given by this analysis (and thus prefer to stick to 'duty' as the logical primitive) he could hardly object to calling the duty unjustifiable.[10]

2 PRESUMPTIVE DEONTIC CRITERIA

Deontologists are no doubt restless with the preceding account, especially with respect to the last part—the proposal for justifying the obligation-making institutions themselves. And while the axiological route just outlined avoids some of the standard objections from deontologists, one would be less than candid if one did not admit that its emphasis on the values involved inevitably dilutes the emphasis on 'being bound' to a course of conduct which we sometimes find it important to convey. There are some obligations, a deontologist will argue, which are simply misrepresented by arguing that they are to be justified *only* by an appeal to the values of the corresponding institutions. He is not likely to dispute that these obligations *can* be recommended on such a basis.[11] He will simply argue for a more direct justification as well. There is a procedure available for making such an argument.

Specifically: when the merits of various obligation-making arrangements (promising, contractual arrangements of various sorts, one's 'station and its duties' arrangements) are under consideration, each of the three features of life used as value criteria operate to give preference to some of these arrangements over others. The purposive, personal, and aesthetic features of human life operate directly in such cases to accept or resist the various alternatives. They can function, then, as presumptive deontic criteria.

To explain: as has been previously remarked, the usual metaphors for explicating the 'sense of obligation,' the 'sense of duty,' and the notion of 'being obligated' are those of *binding*, *being tied to*, or *bound by*. Such locutions as 'duty bound' and 'legally binding contract' are commonplace. But metaphors concerning *boundaries* and boundary-making are even better for getting at the nature of obligation, for while they imply constraint, just as the notion of binding does, they also convey more accurately the central feature of being obligated: that one's conduct options have been limited in certain situations—that one's range of choices has been *bounded*. Such locutions as, 'I had no choice but to do my duty' focus explicitly on this. The function of the practice of fastening people with obligations (from which, naturally enough, some of its values derive) is to proscribe their future conduct—to guarantee (as much as human efforts at reaching agreements and/or using the various forms of coercion can guarantee) that from among the large number of choices people *might* make, they *will* make only specified ones.

On the basis of such guarantees we build most of what we call 'the social order,' and it is obvious that the question of the *value* of the whole practice, as well as of specific instances of it, can be raised immediately. Procedures for dealing with such value questions have been discussed. But here it is important to note the way in which the features of life used previously as presumptive *value* criteria also operate to select and rank-order various obligation-making arrangements—and thus can function, at least in disputes over which arrangements to choose, as presumptive *deontic* criteria.

PERSONALNESS

It is convenient to begin with the personalness feature, for its function here is the clearest model for the way the other two function. It has been noted earlier that the notion of 'personhood' has a good deal to do with the awareness of a boundary between self and 'other.' The earlier discussion also pointed out that violations of one's person—events involving trespass of the boundaries in some way—are resisted. The point to be added here is simply that the impulses one has to resist violations of his person are quickly joined by insistence that such violations do not occur in the future. That is, in the normal course of maturation, one soon extends one's concerns for one's person from resisting actual violations to *prospective* concern—to insistence that such violations do not occur again. It is at this point that our' personalness' begins to operate as a criterion for ranking various obligations and obligation-making arrangements.

For what is it to insist that others do not violate one's person if it is not to insist that they conduct themselves within certain 'bounds'? In so far as one insists on protecting one's person from violation, does one not have built-in (operative) preferences for arrangements which attempt to guarantee inviolateness as against ones which do not? And do we not have an operative preference for arrangements which protect us against some *sorts* of violation (e.g., capricious and unjustifiable violations) as opposed to other sorts (e.g., systematically justifiable ones)?

The argument for the justifiability of the use of the personalness feature as a presumptive deontic criterion is the same as the argument for its use as a presumptive value criterion. Namely, it is a set of priorities which simply goes into operation prior to any 'moral choices' one might make and independently of any culture-bound socialization process one might undergo. And if one can find no reason to change, then continuing with these priorities cannot be, from a reasoned point of view, *wrong*.

Of course, such guidance as is given by this presumptive criterion is only a small piece in the whole process of justification. Many important questions get no help at all from this source: which of the 'protective' obligation-making arrangements is the best (i.e., most efficient)? That is apparently a question to be handled in terms of the values involved. Should one use the practice of assigning and enforcing obligations *at all* in attempting to guarantee future conduct? That too seems handleable only in an axiological or agent-morality framework. Further, the preferences operative here may conflict with those from the purposiveness or aesthetic features of our lives. If so, there is no 'presumptive guidance' given.

But some guidance is better than none, and when combined not only with the priorities operative in the other criteria, but also with the axiological scheme and the agent morality considerations to be developed later, will not seem so inadequate after all.

PURPOSIVENESS

Whatever value priorities are operative in the purposiveness of life also translate directly into presumptive criteria for dealing with conflicts among various obligation-making arrangements. The 'prospective' nature of purposive conduct is obvious: goal-seeking usually requires planning, and planning depends on expectations about the future. The more one can count on the future being a certain way, the more effectively he can plan. Obligation-making arrangements, as ways of guaranteeing some things about the future, obviously have consequences for purposiveness, and the value priorities here translate directly into preferences for arrangements which attempt to guarantee success for purposive action in line with those priorities. Thus, confronted with a choice between arrangements designed to guarantee the success of purposes whose satisfactions are long lasting, intense, and fecund, it is not surprising to find an operative preference for the latter.

Again, the argument for the legitimacy of the use of these operative priorities as *presumptive* criteria is the same as in the section on value criteria. The possibility of conflicting priorities from the other criteria is acknowledged. The guidance given by this one criterion alone is only a small portion of an adequate set of justification procedures.

THE AESTHETIC NATURE OF LIFE

The argument here is parallel to the preceding two. The moment that prospective considerations enter the picture, all the value priorities mentioned in the discussion of presumptive value criteria become presumptive deontic criteria—operative preferences for those obligation-making arrangements which attempt to guarantee conduct from others which will not need an 'avoidance reaction,' or which will get an 'approach reaction.'

The legitimacy of using these operative priorities as presumptive deontic criteria is based on the same argument for presumptiveness used in the section on value criteria. And again, the fact that the guidance provided here is both very partial and subject to cancellation from conflicting priorities in the other criteria is acknowledged.

3 PUTTING THE TWO PROCEDURES TOGETHER

To summarize, then: it may be possible to ground the justification of obligation-statements—without exclusive reliance on axiological considerations—in the following way. After showing that the assertion of an obligation is entailed by some obligation-making institution (Searle's procedure), one raises the question of the appropriateness of the institution itself. There are several aspects to this question, and some of them undoubtedly concern the values involved. For example, one may want to know whether the whole practice of erecting obligation-making institutions is effective for guaranteeing things about the future. One may want to know which, among several possible ways of defining a particular sort of institution, is the most 'fitting,' most efficient, or most pleasing. These are obviously questions of value—for which the procedures developed in earlier sections are appropriate. But in the process of choosing among alternative definitions of an obligation-making institution (e.g., choosing how rigidly to define one's obligation to keep promises), one may well find a presumptive ground for one's inquiry in the operative priorities of the deontic criteria. The justification of obligation-making institutions (and thus of obligation statements), while not separable from considerations of value, is thus not limited *only* to procedures in an axiological scheme. And that, indeed, fits nicely with our common sense hunches about the matter.[12] There are some bothersome sources of uneasiness, however.

It may be felt, for example, that a virulent form of ethical egoism has been introduced in the description of presumptive deontic criteria. For all the priorities mentioned seem to relate only to others' conduct—making sure others don't violate one's person, and that others don't interfere with one's purposive behavior or pleasure. Where is the warrant for one's *own* obligations? Are we to conclude that the deontic criteria would give equal weight to both universalized obligation-making arrangements and arrangements which except oneself?

The answer to such questions must be given in several parts. First, it must be remarked that to the degree that an obligation on *others* can be grounded by reference to one's own purposiveness or personalness, to that degree an obligation *on oneself* can usually be grounded by reference to *others'* purposiveness or personalness. And there is certainly nothing paradoxical about saying that the obligation-making institution which binds me not to violate another's person is grounded in *his* operative priorities rather than my own. Indeed, this is precisely what the deontologist wants to insist on—that in a most fundamental sense, my acts of obligation and duty are *for* others, are not to be justified in terms of *my* values, or *my* priorities, but in terms of *theirs*.[13]

But why consider their priorities at all in the process of deciding the merits of various obligation-making arrangements? Well, the situation here is precisely the same as it is in evaluating conduct. There it was argued that others have values for my conduct, and those values are every bit as 'real' as my own values for my conduct. If I am interested in a *reasoned* evaluation of my conduct, I cannot omit others' valuations from any considerations *without reasons*. To do so would be arbitrary and thus inconsistent with the project underway. Just so in the case of the operation of presumptive deontic criteria: whether the presumption in favor of *my* being bound comes from *my* personalness or from someone else's is irrelevant. It is, nonetheless, a presumption in favor of my being bound. Unless there is some reason for disregarding it, its use in deciding the case cannot be, from a reasoned point of view, wrong.

There are, of course, two sorts of reasons which might be offered to confound the presumptive validity of justification in terms of deontic criteria:[14] conflicting presumptions and countervailing values. When an obligation-making institution is acceptable in terms of one person's criteria ('No skin off my nose!'), but is rejected in terms of another's (e.g., as a violation of his person), the conflict nullifies the whole presumptive procedure and one's usual recourse is to argue in terms of the balance of values. This is, in fact, the way we ordinarily proceed in cases where self-sacrifice is involved.

The second sort of reason one can offer to confound the use of presumptive deontic criteria is of course the citation of countervailing values. One can occasionally find that an evaluation of some obligation-making arrangement leads to the conclusion that one ought *not* to have it, while in terms of the presumptive deontic criteria the arrangement is operatively preferred. The age-old conflicts of duty and interest exemplify these situations well. And it will not do to dismiss such cases lightly. At best they can be handled by finding that the balance of values favors *some* obligation-making institution consistent with presumptive deontic criteria—even though the values for any given individual in such a situation are overwhelmingly bad—and even though it is not the 'highest' priority in terms of the deontic criteria. Examples here are easily drawn from the compromises we have to face in our crowded society: e.g., population control. Some arrangement for preserving one's person and purposing against the interference the sheer size of the population can cause is surely accepted in terms of the deontic criteria. But several sorts of arrangements are possible, some which have severe disvalue for large numbers of people. The trick is to find one which is both consistent with the operative priorities of the deontic criteria and which, at the same time, can be justified in terms of the balance of values involved. That is the best one can do with such conflicts. At worst, and more often than is comfortable, such conflicts become an impasse in which all presumptive guidance is lost.

But moral scepticism cannot survive on occasional impasses—no matter how poignant they are. It needs an unrelieved diet of blocked justifications. That it plainly cannot have if these arguments stand.

NOTES

1 It is not maintained that the two uses of 'ought' put forward in the arguments of 'From Is to Ought' exhaust the analysis of that word. Indeed, several more uses have been noted. See: H. N. Castañeda, 'Imperatives, Oughts, and Moral Oughts,' *Australasian Journal of Philosophy*, 44 (1966), pp. 277–300; J. Ward Smith, 'Impossibility and Morals,' *Mind*, 70 (1961), pp. 362–75; W. J. Rees, 'Moral Rules and the Analysis of Ought,' *Philosophical Review*, 62 (1953), pp. 23–40; B. J. Diggs, 'A Technical Ought,' *Mind*, 69 (1960), pp. 301–17; Peter Glassen, 'The Senses of "Ought",' *Philosophical Studies*, 11 (1960), pp. 10–16. Two uses are enough for present purposes. It should be borne in mind, however, that the second usage by itself—the so-called 'moral ought'—is not enough. To confine the discussion of ought statements to obligation assertions, and further to suggest, even if uninten-

tionally, that only this sense of 'ought' is of moral concern, is to open wide the gates of misunderstanding. Such abbreviations of the usage of 'ought' are not uncommon. See, on this point, Jonathan Harrison, 'Moral Talking and Moral Living,' *Philosophy*, 38 (1963), pp. 315–28; and for an example, A. C. Ewing, 'The Possibility of an Agreed Ethics,' *Philosophy*, 21 (1946), pp. 29–41.

2 See the extensive commentary on 'ought' in the *Oxford English Dictionary* on this point and the ones to follow.

3 Note here that when one can offer a reason against 'urging' *x*—against moving from right to ought statements about it—one may say that in this case he has a reason for concluding that it is not the case that he ought to do what is right. Care must be taken not to let 'ought' carry the freight of its 'obligation-sense,' of course. And, 'It is not the case that one ought to do what is right' is no doubt best distinguished from, 'One ought not (in this case) to do what is right,' though one has a suspicion that in ordinary parlance the interchange is frequently made. For a treatment of 'ought' with interesting similarities, see Bertocci, 'The Authority of Ethical Ideals,' *Journal of Philosophy*, 33 (1936), pp. 269–74. This article expresses the 'emphasis' point nicely, but seems to conflate the two senses of 'ought' I distinguish.

4 John R. Searle, 'How to Derive "Ought" from "Is",' *Philosophical Review*, 73 (1964), pp. 43–58. The procedures outlined here may not be at all points what Searle had in mind. No attempt has been made to stay with his exposition. But much of it has been lifted directly from his article, and all of it was suggested by his article. One should also repeat here Searle's acknowledgment of G. E. M. Anscombe's article, 'On Brute Fact,' *Analysis*, 18 (1958), pp. 69–72. Also of help was John Rawls, 'Two Concepts of Rules,' *Philosophical Review*, 64 (1955), 3–32. The reader may also want to consult the critique of Searle's article by James and Judith Thompson, 'How Not to Derive "Ought" from "Is",' *Philosophical Review*, 73 (1964), pp. 512–16. Also, Max Black, 'The Gap Between "Is" and "Should",' *Philosophical Review*, 73 (1964), pp. 165–81; A. N. Prior, 'The Autonomy of Ethics,' *Australasian Journal of Philosophy*, 38 (1960), pp. 199–206; and D. W. Hamlyn, 'The Obligation to Keep a Promise,' *Proceedings of the Aristotelian Society*, 62 (1961–2), pp. 179–94.

5 Beccaria says: 'The word "obligation" is one of those that occur much more frequently in ethics than in any other science, and which are the abbreviated symbol of a rational argument and not of an idea. Seek an adequate idea of the word "obligation" and you will fail to find it; reason about it and you will both understand yourself and be understood by others.' (*On Crimes and Punishments*, trans. Henry Paolucci, New York: Bobbs-Merrill, no date; footnote, p. 15.) The analysis of 'obligation' given here certainly agrees with Beccaria's view in one respect: the notion of an obligation is not straightforwardly definable in a phrase or even a few phrases. It is rather an abbreviation for a complex state of affairs in which one's options are bounded by the terms of some institution which entails judgments about what it is right to do (and thus what one ought to do in sense (1) of 'ought'). The 'sense of obligation' is, one supposes, the awareness of such a situation.

6 The locution 'the obligation sense of ought statements comes from the notion of bounding a man's choices by certain institutions. . .' was chosen

with a good deal of care. In the first place, the usual metaphors for the explication of obligations are those surrounding the notion of binding, being tied to or bound by. It is felt that the notion of one's options being bounded was a more revealing metaphor. Secondly, sometimes one carelessly says that the obligation comes from binding oneself, or placing oneself under the system of rules. This is misleading. We often want to say that a man has been placed under an obligation by circumstance, or simply has an obligation, where there is no suggestion that he has, or need have, consented to anything. So it is best to keep the locution 'impersonal'—the obligation comes from the bounding of a man's choices, not his bounding of his choices. Further, of course, if promising is construed as an obligation-making institution, there is more than a hint of circularity in saying that a man 'binds' himself to keep a promise.

7 Hubert Schwyzer, 'Rules and Practices,' *Philosophical Review*, 78 (1969), pp. 451–67.

8 Searle, *op. cit.*, p. 57.

9 See, for example, David P. Gauthier, *Practical Reasoning* (Oxford: Clarendon Press, 1963).

10 Compare the arguments of W. D. Falk in 'Obligation and Rightness,' *Philosophy*, 19 (1944), pp. 129–47.

11 Deontologists have never wanted to argue, I think, that considerations of value are always *irrelevant* with respect to deciding what one's obligations are—only that they are not *always* relevant, and not always decisive when they *are* relevant. On the present analysis, the value questions *are* always relevant at some point (e.g., at the level of using any obligation-making institutions at all)—unless one wants to treat something like Singer's Generalization Principle (treat similar persons in similar circumstances similarly) as an 'obligation.' I argue that it is instead simply a formal part of the reasoning process. (See below, Chapter XII.) So it is hard for me to conclude that axiological considerations are ever irrelevant to the justification of obligation statements. But neither are they the *whole* of the justification process. In those two conclusions lies the contention about the co-ordinate status of axiological and deontological approaches. Agent-morality views will be brought in later.

12 The remark that a co-ordinate status for axiology and deontology fits with our common sense hunches is simply a reminder of the arguments in Chapter II to the effect that people are typically uncomfortable about both disassociating questions of duty and value on the one hand, and completely subordinating the former to the latter on the other hand. Common sense may not be correct, but in the absence of convincing reasons to doubt it, the 'test of time and practice' is the most reassuring of the appeals to 'authority.'

13 The fact that the presumptive validity of one's duties and obligations is often to be argued by reference to operative priorities *others* hold for one's conduct (which priorities, even if they do not *conflict* with one's own, still *are* not 'one's own') helps to explain the sense of locutions like, 'The duties *imposed* on me. . .' and the conviction we often attend to that duties, obligations, and responsibilities are essentially independent of us—in the sense that they devolve 'upon' us 'from' various things which are 'outside' ourselves

and not necessarily related to our own needs, desires, or priorities. And it explains, at least to me, why my reaction to an unrelieved deontological ethic, and even more to an unrelieved use of a deontological approach in the moral life, is one of alienation.

14 When agent morality considerations are brought in, there will be three.

XI

THREE ISSUES CONCERNING THE ARGUMENTS SO FAR

Suppose, now, that we have procedures adequate for (a) valuing a course of conduct; (b) evaluating that course of conduct; (c) grounding evaluations; (d) getting from is to ought—in two senses of 'ought'; and (e) grounding justifications of obligation-statements, both by way of an axiological scheme and by reference to presumptive deontic criteria. Many issues remain as yet untouched: e.g. the ascription of an obligation to an individual, holding him responsible for fulfilling it, taking steps to see that he does, taking steps when he does not—all these issues raise special problems. And this is not even to mention the difficulties peculiar to agent morality considerations.

But before embarking on these other matters, a few features of all the arguments so far outlined require discussion.

1 THE 'CETERIS PARIBUS' CLAUSE

The first thing to note is the presence, in all the argument forms of Chapter X, of what has been called the *ceteris paribus* clause—or at any rate, the presence of a premise equivalent to it in effect. One argues that x is right because it is good, and 'bad in no countervailing sense.' This is to say, in effect, x is good, and 'other things being equal,' it is right to do. And so on with the arguments leading to conclusions that one ought to do x. Each of these arguments will contain a deliberate *ceteris paribus* disclaimer: the conclusion goes through unless there are countervailing reasons to the contrary. This is not unusual in any way; all arguments may be construed to contain such a disclaimer covertly. The very use of evidence to establish a conclusion entails the consequence that evidence to the contrary would have the reverse effect. Any attempt to suppress consideration of such counter-evidence from judgment of the conclusiveness of the argument commits the fallacy of special pleading.

But making the *ceteris paribus* clause explicit in the procedure just outlined has some interesting by-products. For one thing, it makes obvious and understandable the difficulty of arguing about hypothetical moral cases. This has been mentioned before but deserves a bit of elaboration. The *ceteris paribus* clause reads, 'If there *are* no reasons to the contrary' ('If x is in no countervailing sense bad,' etc.), and in the context of an actual case, the premise may not be overwhelmingly difficult to prove. Granted it is a negative universal, apparently requiring a survey of the entire range of possibility for its proof, but we have common in-use procedures for dealing with such premises (think of, 'There are no elephants in this county'), and with the requisite nod to epistemologists, one should not hesitate to call the result of such procedures 'reasoned argument,' or 'warranted conclusion,' or 'proof.'

But this applies to actual cases, where wild speculations ('But of course some elephants could have migrated over the polar cap, carrying their own food.') can be turned aside by showing that there is no reason to believe such an event occurred, and thus it has no force for the argument at hand. In hypothetical cases, however, the number of 'reasons to the contrary' to be found—and thus the adjustments to be made to accommodate them—are limited only by one's imagination, and the cunning with which the case description can be hedged. Concentrating on the way hypothetical cases must be hedged to make arguments work creates the suspicion that a rational defense of a moral judgment is possible *only* in the realm of hypothesis, where whole regions of embarrassing possibilities can arbitrarily be defined out to allow the arguments to go through. This suspicion is nearly in direct opposition to the situation described in these pages: namely that it is in most cases more difficult to deal with the *ceteris paribus* clause in hypothesis than in practice.

A further feature of the procedures outlined so far that is made evident by the *ceteris paribus* clause is how thoroughly contingent justified judgments are. If circumstances change, the *ceteris paribus* premise may turn truth values, and an argument which once went through may no longer do so. The consequence in practice is that one must develop criteria for deciding what sorts of circumstantial changes could conceivably change matters and reassess the status of the relevant moral judgments whenever such changes occur. This does not usually mean chaos at all—especially for the broader moral principles whose warrant is ordinarily to be found in rather persistent, if not permanent, features of our lives and environment. It only means that no moral judgment is any more permanent than its rationale—which is precisely the situation one wants if he is interested in a reasoned approach to these matters.

2 UNIVERSALIZABILITY

Of course, the sort of contingency under discussion here has nothing to do with whether or not a moral principle 'established' by these means is or is not universalizable or 'categorical.' On the matter of universalizability: any assertion of what one ought to do may be regarded as a rule of conduct when it is tautologous to infer that, 'If one ought to do A, then similar persons in similar circumstances ought to do A.' This inference is tautologous if one has arrived at a decision on what ought to be done by the procedures here outlined, for one arrives at that decision by considering 'circumstances.' Obviously, similar enough circumstances will produce the same decision (given the same decision procedure). In some cases one is inclined to think that this 'similar persons in similar circumstances' clause reduces to saying, 'Anyone who is person *x* here and now ought to do A.' One is tempted to think this from the suspicion that some circumstances and personal abilities are so unique as to prohibit repetition (in the case of circumstances) or duplication (in the case of personal abilities). If so, then in those cases the 'rule' goes no further than 'this one here and now.'

If one asks whether every assertion about what one ought to do 'must contain' the general assertion about similar persons in similar circumstances, the answer is no. That is, as a question simply about the possibilities of language-use, the answer must be that a man might very well utter a judgment about what he (or some other fellow or group) ought to do which excludes this generalization. After all, it is possible to get the idea of such an ungeneralized idea across or one could not talk about it. And if it is possible to get it across by explicit reference to it, then one sees no reason that someone could not get it across (mean it and have his meaning understood) by the phrase 'one ought to do A.' One sees no reason to suppose that the phrase must always contain the generalization as part of its meaning. In fact, with our usual tendency to make exceptions for ourselves—and only for ourselves—in some matters of conduct, this ungeneralized assertion may be more common than we would like to think.

Ordinarily, we find fault with an ungeneralizable judgment about what one ought to do. As noted in Chapter IX, one may even insist that ungeneralizable judgments are non-moral. Certainly, such fault finding is an inevitable consequence of insisting on reasoned conclusions about what to do. As noted above, however, any 'ought' judgment produced by the procedures here outlined is universal. That is, any reasoned conclusion will hold whenever those reasons occur. Unless the reasons are limited by circumstance to apply to only one individual, the conclusion will hold for similar persons in similar circumstances.

As for the issues raised by the distinction between categorical and hypothetical principles, the procedures outlined here do not prejudice the outcomes one way or another. Kant's categorical imperative, at least on one interpretation, looks very much like a restatement of the formal feature of reasoning just noted in the discussion of generalizability, however.

3 EVIDENCE TO THE CONTRARY

To return to the *ceteris paribus* clause: what remains to be discussed are some general types of evidence which the *ceteris paribus* clause is designed to 'cover'—that is, the sort of 'other things' which one wants to be 'equal' in order to get an argument to go through. What sort of contentions, then, might be advanced in one case or another to challenge the claim that 'other things are equal'?

Well, of course, there is 'countervailing valuation' in the case of premises asserting a particular value for a conduct option. And dispute of the claim to have gotten a fair balance (so as to decide that a course of conduct is 'right'). These must be handled by the procedures heretofore discussed for valuation and evaluation.

There is also (in the case of 'ought statements') the agency-excuse—i.e., the claim that to say A (a particular individual) ought to do *x* (even though the general rule, 'One ought to do *x*' is admitted to be justifiable) is false due to the fact that A is not a 'responsible agent.' This sort of claim is best dealt with later, in the context of the fuller discussion of the concept of responsibility.

A third form of *ceteris paribus* evidence is something occasionally dubbed the generalization argument.[1] It argues against the 'rightness' or the obligatoriness of an act by calling attention to the consequences of everyone's doing it. It is commonly heard in negative form: 'You can't do that. What if everybody did that? The world would be in a mess then!' In its general form, it goes: it is sometimes relevant to consider what would happen if a piece of behavior were generally engaged in, in deciding whether one ought to do it.

One wants to say that the general form goes 'It is *sometimes* relevant. . .' for most everyone recognizes that it is not always relevant. If everyone were to be a philosopher, for example, the world would starve. But since very few are so inclined, that fact need not keep me from deciding to be a philosopher. The same is true of nearly every profession: if everyone were a farmer, the world would lack medical care. So the argument must be put as, 'It is *sometimes* relevant. . .' The issue to be decided is how and when the generalization argument is relevant to reasoning out what one ought to do.

In outline, the procedure is as follows: The relevance of the generalization argument is directly proportional to (a) how much generalizing the behavior in question would affect the balance between its goods and bads; and (b) the probability of its becoming generalized.

(a) If generalizing the behavior in question either does not alter the balance of goods and bads or else changes the balance by adding to the goods, the generalization argument can hardly be relevant to a prohibition of that behavior. If the generalization of the behavior swings the balance over to bad (or even increases the bad) then it may be relevant to a prohibition. If it increases the good, it may be relevant to the recommendation of the behavior. If it does not affect the balance at all, then it is not relevant at all.

The degree of generalization is also a relevant consideration. If one can expect an increase of 10 per cent, but no more, and 10 per cent is desirable (although 100 per cent would mean disaster), then the relevance of the generalization argument is somewhat different than if one could expect the 100 per cent increase.

(b) In order to secure the argument's relevance, however, another consideration enters. If there is very little (or no) probability of the actual generalization of the behavior, then the argument fails, to that degree, to have force for actual cases.

Procedurally, this is as far as one can go with the generalization argument. In practice, its use is usually quite disputable, owing to the 'looseness' of procedures for determining relevance. It usually functions, however, as an *auxiliary* argument to shore up otherwise shaky judgments. As such, its 'looseness' need not give one too much concern. If it can be made to work, fine. If not, all is not yet lost.

Finally, there are those 'infelicities' mentioned earlier which provide grounds for disputing the 'one ought to do *x*' conclusion. These apply, of course, to the obligation sense of 'ought' and have to do with the exact way in which the obligation-making institution has been defined. ('But the license was a phoney; they are not married; he has no obligation. . .') No further procedural matters remain here. The problems lie with getting and justifying particular definitions for institutions, or in discovering, in each case, precisely what the definitions are and what they entail. These are not questions of procedure but execution.

NOTES

1 See, for an elaborate discussion of the generalization argument, Marcus Singer, *Generalization in Ethics* (New York: Alfred A. Knopf, 1961). Singer attempts to use this argument as one of the fundamentals in providing a rational basis for ethics, and an impressive volume of critical literature has developed with regard to his attempt. See, for example: Robert L. Holmes, 'On Generalization,' *Journal of Philosophy*, 60 (1963), pp. 317–23; David Keyt, 'Singer's Generalization Argument,' *Philosophical Review*, 72 (1963), pp. 466–76; H. J. McClosky, 'Suppose Everyone Did the Same,' *Mind*, 75 (1966), pp. 432–3; George Nakhnikian, 'Generalization in Ethics,' *Review of Metaphysics*, 17 (1963–4), pp. 436–61; and Warner Wick, 'Generalization and the Basis of Ethics,' *Ethics*, 72 (1961–2), pp. 288–98. See also, A. K. Stout, ' "But Suppose Everyone Did the Same",' *Australasian Journal of Philosophy*, 32

(1954), pp. 1–29; and A. C. Ewing, 'What Would Happen if Everybody Acted Like Me?' *Philosophy*, 28 (1953), pp. 16–29. On the matter of universalizability—what Singer calls the generalization *principle*—see two articles by Alan Gewirth: 'Categorial Consistency in Ethics,' *Philosophical Quarterly*, 17 (1967), pp. 289–99; and 'The Generalization Principle,' *Philosophical Review*, 73 (1964), pp. 229–42.

XII

WHY BE MORAL?

We come now to the general question, 'Why be moral?' A good deal of attention has been given to it,[1] but the treatment here will not attempt to relate itself to that work, even though it undoubtedly bears great similarity to some of it and acknowledges indebtedness to it all. Rather the procedure here will be to show how the results of earlier chapters help to dispatch the usual concerns indicated by the question in a way that neither violates common sense nor provides an impediment to the grounding of moral judgments.

The question, 'Why be moral?' actually resolves into a wreath of related questions. They are best dealt with when carefully separated. For example, if the question is taken merely as a demand for reasons for doing what is right and of moral concern, it is obviously senseless in the context of the procedures outlined here. In the case of some other set of procedures for deciding what is right (tarot cards or some such thing), the question might be applicable. But here, one decides what is right by deciding what conduct one can give reasons for. So to conclude that *x* is right —on the methods outlined here—is to conclude that it is a course of conduct which one has reason to do. The demand for reasons for doing it is redundant.

One might insist, however, that the point at issue is not the demand for reasons for doing what is right, but reasons for doing those right things we call moral (as opposed to non-moral things). That is, it might be insisted that 'Why be moral?' means why opt for the morally right over some other kind? The answer must be that it looks very much as if what is morally right, as opposed to non-morally right, is *defined* in terms of what one has reasons to regard as more important and thus over-riding. So the reasons asked for have been given in the very process of making the distinction: that giving such reasons (to decide what sorts of right conduct are remarkably more important than 'run of the mill' ones) is what one means to do when he makes the distinction between the moral and the non-moral. The demand for reasons for doing what is moral, then, is as redundant as the demand for reasons for doing what is right.

It should be noted of course that on some other analyses of these matters—some other procedures for deciding what is right and moral—the demand for reasons might not be redundant. If what is right and moral is decided, for example, by some purely arbitrary or conventional procedure, then the demand for reasons for opting for the morally right over other kinds is not at all out of place.[2]

If the question is rephrased again, however, so as to ask, 'Why *ought* I to do what is (morally) right?', the answer here must be, in effect, 'Why not?' If it is sense (1) of 'ought' that is being used, the reader will recall that all this does is to import emphasis or a 'sense of urgency 'into the judgment that *x* is right, and the procedure was to let such an import pass if no reasons could be given against its use. So the reply to the demand for reasons, here, is 'Why not?' The issue of whether or not it is a moral rather than non-moral conduct option is immaterial; the question may be asked of either, and is answered in the same way in each case.

But if the 'ought' is taken in sense (2), the question becomes, 'Why is one obligated to do what is (morally) right?' Or, 'Why is it one's duty or responsibility to do what is (morally) right?' And this is a nice question. For unless one can find some obligation-making institution behind the whole moral enterprise—the 'moral point of view'—one does not readily see how one can claim to have such an obligation. The question is, again, a demand for reasons for being obligated to do what is moral, and the only procedure advanced in these pages for giving a reasoned justification of an obligation is by way of obligation-making institutions. If, however, the moral enterprise as a whole—in particular, for purposes of answering this question, the business of sorting conduct, principles, and values into moral and non-moral categories, the moral being overriding—is shown to be obligatory when one's options are bounded by a more fundamental institution, *and one can secure reasons for bounding people's options by this institution*, then the procedures are straightforward for arguing that one has obligations to do what is morally right.

Now it may be objected that 'obligation' is a moral category, and that there are no non-moral obligations. And so to say one has an obligation to be moral is an absurdity (circular, or involving a paradoxical self-reference). It would be like saying one has a duty to do one's duties, a responsibility to carry out one's responsibilities. And what could one mean by such assertions in addition to the obvious emphasis they give to the assertion that one simply *has* obligations, or duties, or responsibilities? Is not the assertion of a duty to do one's duty an empty one? And thus the demand for some reason to suppose one *has* such a duty a meaningless question? The objection is a clever one, but it will hardly bother the persistent questioner, who will merely rephrase his question, avoiding the term 'obligation.' He will simply ask, then, what reasons can be given for getting involved in the whole business of having obligations, duties, and responsibilities in the first place. But the reply procedure outlined in the preceding paragraph can easily be adapted to the rephrasing. The point is that procedures exist for giving reasons for adopting 'the moral point of view.' And if these reasons must be considered to be non-moral, and the bounding of one's options to moral ones (when there is a conflict) must be considered a non-moral act in order to avoid paradoxes in discourse, this should cause no problem. It may be necessary to make an adjustment in the analysis of the distinction between the moral and the non-moral, for the bounding of one's options to moral ones is clearly going to meet the criteria for being a moral act in terms of its importance. But resolving the paradoxes of self-reference has often involved the making of *ad hoc* adjustments. The ones necessary in this case are not outlandish.

Now suppose the question is escalated one final time: suppose instead of asking for reasons for doing what is *moral*, or what one *ought* to do, one asks for a defense of reasoning out answers to these questions? This question is, strictly speaking, circular: it asks for reasons in support of reasoning. Thus one cannot answer, strictly speaking, because nothing intelligible has been asked. But the question does not die so easily, and a look at the probable impulse behind it shows why.

It may be suspected that the arguments of this book—indeed, of most of moral philosophy—contain an implicit recommendation that one 'should' engage in valuing and evaluating conduct, and that he 'should' reason such matters out as far as possible. What is wanted is some support for this implicit recommendation, for if a recommendation is made for these options, there might have been others which were not recommended; if a choice was made between the various options, some principle of selection must have been behind the choice; one simply wants to see that principle of choice out in the open. That is surely a reasonable request.

The request will be answered here by noting that, from the mere existence of those features of life that have been called presumptive value criteria and presumptive deontic criteria, there is a great deal of existing and unchosen momentum in the direction of finding the right thing to do by reasoning. In fact, these features presumptively *define* what is right and to have them *in operation*—to have any presumptive value criteria in operation as 'givens' of our lives—is, by that fact alone, to be involved with the business of valuing and evaluating. The criteria themselves are not recommended here; they simply are noted as given and used as long as no reason to the contrary can be secured. And so, *a fortiori*, the same is true of the inevitable entanglement of these 'givens' with valuing and evaluating: unless there is some reason to stop, there can be no reasoned objection to continuing. Similarly for reasoning out answers: this procedure is learned as a way of getting results that work and last and produce agreements among people. As such it may be considered as a product of the momentum generated by the 'givens' mentioned above, and which may be used unless some reason can be given against it. The curious feature here is that while asking if there are any reasons to get into the business of reasoning is circular, asking if there are any reasons for getting out of it is not. And as a matter of fact, there are very often good reasons for getting out. A reasonable man does not always reason.

One final and somewhat different matter: Alan Gewirth has argued[3] that the central meaning of 'being moral' when questions of the 'why be moral?' type are raised, has to do with taking positive account of *other* people's interests. That is, that the central issue here is that of getting out of the (ethical) egocentric predicament. I am inclined to agree. But the answer to the problem stated in this way will be the same: if one is trying to decide what to do by reasoning out an answer, then he will be committed (by the terms of the reasoning procedures developed here) to taking others' interests into account. That does not mean that he will always decide in favor of their interests over his own: for 'his' values may outweigh theirs. In such a case, it would be 'right' and (by this analysis) moral to act in his own interest.

This conclusion is not troublesome at all until agent morality considerations are brought in. For we sometimes want to say that one ought to be the sort of man who is altruistic (or benevolent, or kind, or helpful, etc.) regardless of what it is right to do in terms of the balance of values, or what it is right to do in terms of obligation and duty. 'Why be moral?' then means 'Why be "moral" as a matter of character?' This matter will be discussed in Chapter XIX.

NOTES

1 See, for further discussion of these matters: S. Toulmin, *The Place of Reason in Ethics* (Cambridge: Cambridge University Press, 1960), 14.1 ff.; Kurt Baier, *The Moral Point of View* (second edition, New York: Random House, 1965), Chapter 7; Marcus Singer, *Generalization in Ethics* (New York: Alfred A. Knopf, 1961), pp. 319–27; F. H. Bradley, 'Why Should I be Moral?' in his *Ethical Studies* (reissue of second edn, Oxford: Oxford University Press, 1962); J. S. MacKenzie, 'The Source of Moral Obligation,' *Ethics*, 10 (1899–1900), pp. 464–78; A. I. Melden, 'Why Be Moral?' *Journal of Philosophy*, 45 (1948), pp. 449–56; two pieces by Kai Nielsen, 'On Being Moral,' *Philosophical Studies*, 16 (1965), pp. 1–4, and 'Is "Why Should I be Moral?" an Absurdity?' *Australasian Journal of Philosophy*, 36 (1958), pp. 25–32; and Alan Gewirth, 'Must One Play the Moral Language Game?' *American Philosophical Quarterly*, 7 (1970), pp. 107–18.

2 Or if, for example, one regards matters of moral concern as distinguishable from others by virtue of their content (such that, for example, they would never advocate the expedient thing to do over the kind or sympathetic thing to do), then the question 'why be moral?' is again more pressing. See G. J. Warnock, *The Object of Morality* (London: Methuen, 1971).

3 Alan Gewirth, 'Must One Play the Moral Language Game?' *American Philosophical Quarterly*, 7 (1970), pp. 107–18.

XIII

THE CONCEPT OF RESPONSIBILITY

In Chapter X it was argued that one of the senses of 'one ought to do *x*' was 'one is obligated to do *x*'—i.e., under an obligation to do *x*. As an analysis of the meanings of 'ought,' that account was inadequate, for obviously 'ought' is not always adequately paraphrased in either of the two ways given there. Yet those two ways do point to a crucial division in the usage of 'ought': one set of uses designed to indicate no more than the way in which what is right is urged on us; the other designed to indicate that not only is the rightness of the conduct at issue, but also some further element, which for lack of a better word may be called a deontic element. This deontic element is sometimes spoken of as an obligation, sometimes as a responsibility, sometimes in terms of duty or even honor.

The concept of obligation, like the concepts of duty and honor, is less inclusive than the concept of responsibility. For one thing there are many instances in which it may be said that, although I have a responsibility to do *x*, I am not, strictly, under an obligation to do it, and it is not exactly a duty for me (or a matter of honor). So while obligations and duties (and perhaps some matters of honor, though this is doubtful) may be expressed in talk about responsibilities, the reverse is not always the case. Further, it is the concept of responsibility, rather than the others, which is used to express and explain the notion of agency often required for the fixing of praise or blame, the assignment of obligations and duties, and the actual sanctioning associated with the practice of holding people responsible. For these reasons it is most convenient to discuss the deontic element of ought statements in terms of the concept of responsibility.

As for why the question of responsibility is taken up at this point rather than, say, the question of virtues and vices: it is purely a matter of convenience. No attempt will be made to suggest that the procedures outlined here are logically prior to agent morality matters. It is rather that the procedural matters so far discussed lead naturally to questions of the ascription of responsibility, the justification of punishment, and the achievement of justice. And whereas grounding procedures previously developed suggest a way of handling decisions about what sorts of conduct can be justified as virtues (allowing one to put off dealing with the question for the moment), the questions associated with responsibility seem more pressing.

I RESPONSIBILITY-STATEMENTS

Any statement which asserts or denies that a human being[1] is (has been, will be) responsible or that he has (used to have, will have, etc.) responsibility, I will call a responsibility-statement—R-statement for economy. R-statements are of several significantly different types.

One set of R-statements is characterized by the fact that they charge a man with responsibility for having 'done' something, or with responsibility for not having done something, or with responsibility for something that has happened. Examples:

(a) 'He is responsible for John's death.' This charges a person with having done something—with having caused John's death in some way (perhaps by pulling the trigger, perhaps by paying someone to do it).
(b) 'He is responsible for that uncovered well; that's why the child fell in.' This charges a person with having failed to do something—with responsibility for something he has not done.
(c) 'He is the employer; the accident was in his plant. He is responsible for the medical bills.' This charges a person with responsibility for something that just

happened. It need not assert that he caused it, nor even that he is blameworthy for it.[2]

R-statements of these three kinds will be called, for lack of a better name, attributive R-statements—since they attribute responsibility to people for things which have happened.[3]

A second set of R-statements is characterized by the fact that they charge a man with responsibility for doing something in the future—not for having done it, but for doing it whenever it is 'called' for. Examples:

(a) 'He is responsible for maintaining law and order here' charges a man with responsibility for doing something. If he fails, he can be indicted, but the R-statement above does not indict him; it does not suggest that he has yet done anything at all.

(b) 'Well, it is his responsibility not to drink when he drives' on the face of it charges a man with responsibility for not doing something. It might be argued that these negative responsibilities can be put positively: 'It is his responsibility to decide not to drink.' For present purposes this issue is not important; the negative wordings will be retained when they occur in common speech.

(c) 'He is responsible for making compensation to men injured in his plant' is another assertion of the same kind.

This type of R-statement will be dubbed, again with some apologies, deontic R-statements.

A third set of R-statements asserts or denies that men have a certain 'characteristic' called responsibility—or that they are the kind of people we describe as responsible. There are actually two types of statements meeting this characterization:

(a) One type asserts or denies that a man has a certain trait of character—like trustworthiness, a sense of duty, and so on. Such statements deal with the kind of thing that a character witness might be called to testify to. Example: 'Well, shouldn't you have foreseen this theft? I mean, the man isn't very responsible is he?'

This type of R-statement will be called a responsible-man statement. The use of 'responsible' here is opposed to 'irresponsible.'

(b) The other type asserts or denies that a person is a 'responsible-agent'—that is, that he is the kind of man who is properly held responsible for things; not that he is the man responsible for some particular thing, but that he is a man who is such that he could be held responsible for this, that, or the other thing.

This final type of R-statement will be called a responsible-agent statement. The word 'responsible' here is not opposed to 'irresponsible,' but to 'incompetent.'

2 HOLDING A MAN RESPONSIBLE

An adequate general explication of the several types of R-statements is not an easy task—even though such statements are usually well understood in their contexts. The problem is to distill from the array of actual uses a general idea (type description, perhaps) of what it is in general to attribute responsibility to someone for what he has done, or to ascribe responsibility to someone for doing something in the future, or to say he is a responsible man or a responsible agent. In short, the task is to discover what is being done, in each type of case, when a man is being held responsible. The purpose is, of course, to make clear what has to be proved in order to conclude that a man is responsible, so that the search for proof procedures will have both direction and a criterion.

Again, as was the case with valuation, it is a bit inefficient to begin the inquiry by asking directly what sentences using the word 'responsibility' can mean. For it is clear that they can be used to express things totally unrelated to the subject at hand. The task is first mark off the practice of holding people responsible—to identify and describe the range of acts involved. Once that is done, an explication of the things R-statements are used to express about those acts is illuminating. Again, however, as with valuation, one must be careful not to concentrate too heavily on the 'essential' or common features of the acts in order to insure a unary meaning for R-statements; the common features are not the only ones which require justification.

What, then, are the leading features of holding a man responsible? (Let us leave aside, for the moment only, the difficult question of responsible agency.) It is assumed by what follows that the practice of 'holding' a man responsible is the one wanted for examination rather than the broader 'determining responsibility for the events.' The latter may indicate merely an attempt to find the causes of things, and there is nothing special to moral philosophy in that activity. But there is something special to moral philosophy in the activity called *holding* people responsible. Just what that specialness consists of will be explicated below. The procedures used in arriving at the explication are the same serviceable—if somewhat uncertain—ones described prior to the discussion of valuation. One final preparatory note: the terms 'answerability' and 'accountability,' often used to explicate the practice of holding people responsible, are not used here. The reason is that these terms too easily confuse the responsible agency question with attributions and (deontic) ascriptions of responsibility. As will be shown, the question of agency is not always included in the others.

ATTRIBUTIVE R-STATEMENTS

Taking as a starting point, then, the division marked by attributive R-statements on the one hand, and deontic R-statements on the other, the question is, first, what is going on when one holds a man responsible for something that has happened? The first thing to notice is that when one attributes responsibility to a man, he is necessarily asserting the man's liability for sanction (praise or blame, reward or punishment), or at least liability for a form of interference (e.g., mandatory rehabilitation or reformation; the mandatory making of compensation). 'Liability for sanction or interference' is here understood to mean that the man held responsible is in a position such that unless there are reasons to the contrary, sanctions or interference of some sort can be justified. To prove an attribution of responsibility, then, it is necessary to prove that (for some reason) the man is in fact 'liable for sanction or interference.' Whether he ought to be sanctioned, and in what manner he ought to be sanctioned, are matters requiring further procedures. Not all men who can reasonably be held liable for sanction ought to be sanctioned.

In addition to the liability issue, holding a man responsible (in the attributive sense) sometimes includes the assertion that he caused a wrong (or right), and sometimes it does not. In the case of paying compensation to a factory worker, no 'causal' assertion need be present. So to prove an attribution of responsibility one may or may not have to prove 'cause.'

And of course the question of responsible agency enters into attributions of responsibility very often as a prerequisite. That is, in many cases (but by no means all) it is necessary to show that the one to whom responsibility is attributed is a responsible agent before the attribution may be said to be justified. Leave aside for the moment what it means to say a man is a responsible agent and how one goes about proving it (whether this is a 'defeasible' concept or not, for example). The point here is simply that, on occasion, to hold a man responsible for something that has happened is to presuppose that he is a responsible agent. And so the proof of that attribution will have to include the proof of the presupposition.

DEONTIC R-STATEMENTS

In the case of holding a man responsible for doing (or not doing) something in the future, a somewhat different analysis must be given. The ascription of 'deontic' responsibility sometimes involves the (conditional) assertion of liability for sanction or interference—but not always. If one 'gives' a soldier responsibility for sentry duty in a war zone, there is indeed the assertion of liability involved: if you (soldier) do x, all is well; if you do not do x—specifically, if you do y—you will be liable for sanction. So to prove this sort of deontic ascription is to prove (i.e., give reasons for concluding) that the one held responsible is in a position such that if he does not do x (what he is responsible for doing), and if there are no reasons to the contrary, it is just to sanction him or interfere with him in some way.

However, some ascriptions of deontic responsibility do not include this notion of liability: consider the situation of a group of desperate and dispirited revolutionaries who have lost hope of success. The leader divides responsibility for an impossible attack, but it would be improper to assume he is asserting liability for sanction in the case of failure or even desertion. He might be doing that (depending, would we say, on his fanaticism?), but he may simply be making assignments. To prove the reasonability of this sort of deontic ascription, then, is not to prove the reasonability of liability for sanction for an assigned task, but merely to prove the reasonability of the assignment (something, by the way, included in the proof of liability for an assignment).

As in the case of attributions of responsibility, responsible agency is usually (but not always) presupposed in the ascription of deontic responsibility. And so to prove the ascription reasonable is to prove the presumption. Cases in which one might argue that responsible agency is not a presumption are, for example, cases involving no liability for praise or blame, but only liability for the making of monetary compensation under certain defined conditions. The presumption of responsible agency seems to be very closely tied to ascriptions and attributions involving liability for praise or blame, punishment or reward—an item to be more fully discussed in the sections on responsible agency. With these remarks as a preliminary, we may proceed to a more detailed analysis along the usual lines.

3 AN ANALYSIS OF ATTRIBUTIVE AND DEONTIC R-STATEMENTS[4]

The kinds of 'things' which are called responsibilities in attributive and deontic statements are, by inspection, the following:

People: 'She is my responsibility now.'
Events: 'The 100-yard dash is my responsibility.'
Behavior: 'My responsibility was to take notes.'
Objects: (as distinct from the above items) 'The building is my responsibility.'

The examples given are deontic R-statements, but it is easy to construct attributions along the same lines:

'That witness was your responsibility and you failed.'
'The course was your responsibility, not mine.'
'Putting up storm windows was my responsibility. Sorry.'
'You lost the job because that building was your responsibility and you ignored it.'

These sample sentences are, of course, elliptical—abbreviated ways of making reference to behavior of various kinds. 'She is my responsibility now' does not mean that a *person* is one's responsibility, but that certain *behavior* with respect to that person is his responsibility. Similarly, it is not the race, but running the race which is the responsibility; it is not the building, but keeping the building in good repair which is the responsibility; and so on through the rest of the examples given.

But what does it mean to say that certain *behavior* is one's responsibility? Again, it appears that this is somewhat elliptical—that is, it appears that at least part of the meaning of these statements is to assert a relation in which one stands to the use of sanctions. For example, if the R-statement is, 'It is your responsibility to keep the building in good repair,' part of what is asserted is that you stand to the business of keeping the building in good repair in a relation such that (for instance) if you keep it in good repair, you are paid a salary; if you do not, you are fired. At this level of specificity, there are as many of these relations as there are distinct R-statements. But it is possible to explicate, in more general terms, the kinds of relations asserted by attributive and deontic R-statements (the point being, to get clear about what kinds of relations one has to 'prove' to demonstrate that a man has responsibilities in these senses).

Such explications have often been attempted with the use of synonyms like 'answerability,' and 'accountability.' The word 'liable' is the key one in the explication proposed here, and it should be noted that it will be used in the straightforward dictionary sense of 'being in a position to incur.' When *x* is 'liable for blame,' he is in a position to incur it. It may not be proper to actually blame him, for whether he is liable for it, and whether he ought to be blamed are separate (though closely related) questions. But if responsibility is attributed to *x* for something, part of what is asserted is that he is (by law, for example, or by custom, or by virtue of what is just) in a position to be blamed (or praised). To assert *x*'s liability is not necessarily *all* that is meant by an attribution of responsibility (see below); so 'liability' is not a complete synonym for 'responsibility.' But it is a central part of most attributions and ascriptions of deontic responsibilities.

One must distinguish sharply, however, the sense used here for 'liable' (that is, 'being in a position to incur sanction') from another common meaning for the word—namely, 'being in a position such that sanction is *likely* to be incurred.' If one is a claims manager in a store, one is likely (that is, one probably will) incur plenty of abuse—just and unjust alike. For one reason or another, the occurrence of the sanctioning is likely. What is meant here is not that the occurrence of sanctioning is likely, but that one is in a position such that if there are no conditions which render the occurrence of sanctioning unjustifiable, then that occurrence is justifiable. Whether the occurrence of sanctioning is also *likely* is a separate matter. The problem for some succeeding sections will be to inquire into procedures for determining whether or not a man is 'in a position such that. . .'

Thus attributive R-statements may be understood as presenting the following possibilities of interpretation in so far as they are used in the practice of holding people responsible.

MODEL STATEMENT FORMS:

(a) '*x* is responsible for having done *y*'
(b) '*x* is responsible for having *not* done *y*'
(c) '*x* is responsible for what happened'

Explication of (a):

(a) presupposes usually that *x* is a responsible agent in the sense that one cannot, without contradiction, assert both that *x* is responsible for having done something and that he is not a responsible agent.
(a) asserts that *x* 'did' or 'caused' *y*—usually in some quite ordinary sense of those words.
(a) asserts that *x* stands to *y* in a relation such that *x* is either liable for praise for it, liable for blame, liable for reward, liable for punishment, liable for making compensation, or liable for some other type of interference with his behavior.
(a) may be used to actually praise or blame *x*.

Explication of (b):

(b) presupposes (always, as far as I can tell) that *x* is a responsible agent.
(b) asserts that *x* stands to *y* in a relation such that *x* is either liable for praise for it, liable for blame, liable for reward, liable for punishment, liable for making compen-

sation, or liable for being interfered with.
(b) may be used to actually praise or blame *x*.

Explication of (c):

(c) may or may not presuppose that *x* is a responsible agent (an incompetent who owns a company still pays workers' compensation).
(c) usually asserts that *x* 'caused' *y* in some sense (again except in cases like workmen's compensation).
(c) asserts that *x* stands to *y* in a relation such that *x* is either liable for blame for *y*, liable for praise, liable for reward, liable for punishment, or liable for making compensation.
(c) may be used to actually praise or blame *x*.

Deontic R-statements may be understood in this way:

MODEL STATEMENT FORMS:

'*x* is responsible for doing *y*'; or one might use:
'*x* is responsible for *not* doing *y*'; or:
'*x* has a responsibility to do (or not do) *y*'
The explication would remain the same (with the exception that 'not' italicized in the explication below would be dropped in the explication of '*x* is responsible for not doing *y*').

Explication

statement ordinarily but not always presupposes that *x* is a responsible agent (exception, again, in cases like workmen's compensation)
statement ordinarily but perhaps not always asserts that *x* stands to *y* in a relation such that *if x* does *not* do *y*, he is liable for blame or punishment or for making compensation or for being interfered with; if he does do *y*, he is liable for praise or reward or getting compensation of some kind.

Some of the relations between attributive and deontic R-statements may be understood in this way: the truth of

'*x* is responsible for having done *y*'

(if it asserts not only the causal relation but that x is liable for *blame* for having done y) *implies* the truth of

'x is responsible for *not* doing y.'

If the attributive R-statement asserts only the causal relation, then no deontic R-statement is implied. Now, if an attributive R-statement

'x is responsible for y (what happened)'

asserts that x is liable for making compensation to z for y, then the truth of that attribution implies the truth of

'x is responsible for making compensation (if y occurs).'

If the attributive R-statement

'x is responsible for y'

asserts, on the other hand, that x is liable for *praise* for y, the situation is somewhat different. The attributive statement *may* imply that

'x is responsible for doing y,'

but then again it may not. Rockefeller was responsible (in large part) for the founding of the University of Chicago; he may well be liable for praise for it. But that is not to imply that he had a responsibility for setting up universities. The implication of the deontic R-statement in these cases is contingent on the situation. I think it is clear that no deontic R-statement *necessarily* implies an attributive one.

The relations here presented are ones found by inspecting examples. There may be others, but these are the only ones which bear significantly on the present study.

NOTES

1 Assertions of collective responsibility are here omitted for convenience. The extent of their usefulness is a bit disputable anyway, and in the cases in which they have an agreed usefulness, they present a procedural problem which can easily be handled by adapting the methods to be presented here for

individual responsibilities. Assertions of purely causal responsibility—such as might be applied to inanimate objects as well as humans—are excluded because the issue here is the practice of holding people responsible. See, for discussions of collective responsibility, articles titled 'Collective Responsibility' by D. E. Cooper, *Philosophy*, 43 (1968), pp. 258–68; R. S. Downie, *Philosophy*, 44 (1969), pp. 66–9; and Joel Feinberg, *Journal of Philosophy*, 65 (1968), pp. 674–88.

2 For the examples in which responsibility may be attributed to non-responsible agents for what they have done (but probably not for negligence), see Philip E. Davis, *Moral Duty and Legal Responsibility* (New York: Appleton Century Crofts, 1966), the case of *McGuire vs Almy*, p. 68–72.

3 I am indebted to Alan Gewirth for his prodding on the matter of nomenclature here. My original usage was 'indictive R-statement'—unsatisfactory, obviously, because it is so thoroughly associated with wrongdoing, when what is wanted here is a term suitable for reference to either rightdoing or wrongdoing. 'Attributive' (one of Gewirth's several suggestions) has, perhaps, an unfortunate association with the noun 'attribute.' But in its usage as a verb, it will do. 'Ascription' would also do, but it is saved for another purpose.

4 Regrettably, most analyses of the notion of responsibility, while not wrong as far as they go, are quite partial, dealing mainly with attributions and often assuming that the presupposition of responsible agency is necessary to them all. Abbreviated remarks in close agreement with the present analysis may be found in Joel Feinberg, 'Problematic Responsibility in Law and Morals,' *Philosophical Review*, 71 (1962), pp. 340–51. Also see: H. Fingerette, 'Responsibility,' *Mind*, 75 (1966), pp. 58–74; John Ladd, 'The Ethical Dimensions of the Concept of Action,' *Journal of Philosophy*, 62 (1965), pp. 633–45; and two critiques of Ladd's article—Kurt Baier, 'Acting and Producing,' *Journal of Philosophy*, 62 (1965), pp. 645–8; and J. B. Schneewind, 'Responsibility and Liability,' *Journal of Philosophy*, 62 (1965), pp. 649–50.

The analysis given by H. L. A. Hart in Chapter IX of his *Punishment and Responsibility* (Oxford: Oxford University Press, 1968) is also quite close to the one given here. He distinguishes (a) role responsibility (included in but not equivalent to my category of deontic responsibility); (b) causal responsibility; and (c) legal liability responsibility (which seems to be included, but again not quite coextensive with my category of attributive responsibility). He also lists capacity-responsibility, which is precisely what I call agent-responsibility.

XIV

RESPONSIBLE AGENCY

As the question of responsible agency comes up in the course of most proof procedures for attributions of responsibility and ascriptions of deontic responsibility (henceforth, simply *ascriptions* of responsibility), it is reasonable to turn next to that matter. To establish procedures for deciding the responsible agency question will be to nail down one step in the procedures for justifying both attributions and ascriptions of responsibility.

What does it mean to say that a man is a responsible agent? Philosophers have answered in a variety of ways: that it means the man 'is free to choose his course of action,' that he 'could have done otherwise than he did,' that he has 'response-ability,' that he has the ability to make rational choices and to put his decisions into action.[1] No matter what the answer to the question, it has invariably led the discussion into the morass of the free will controversy—from which it has rarely emerged with enough momentum or reader interest to take on many of the other problems associated with the concept of responsibility.

1 SOME CASES

To get a fresh start, specifically in the direction of the task of finding adequate justification procedures, it is illuminating to consider some cases in which the issue of whether or not a man is a responsible agent must be brought to a decision. Consider the following exchange:

'He killed a man and ought to be punished. That's all there is to it.'

'That's *not* all there is to it. He was temporarily insane; he wasn't responsible at the time.'

'What does that mean—"he wasn't responsible at the time"?'

'It means he couldn't help doing what he did—he was extremely upset, unbalanced, couldn't think straight. He didn't realize what he was doing. He wasn't responsible at the time.'

Another case:

'When a man's wife is attacked. I say that man has a moral responsibility to defend her. Smith didn't.'

'But surely you know that Smith is an invalid. He can't even get out of his bed. He is not responsible in those kinds of situations. You can't hold a man responsible for something that it is physically impossible for him to do.'

A further example:

'That janitor stood around doing nothing while the steam built up to the point of explosion. I know it isn't specifically his job to watch the pressure, but he is still guilty of negligence, and I hold him responsible for what happened.'

'But in order for him to *understand* that there was any danger in what was happening, he would have had to have been able to read the gauges and make certain simple calculations. You *know* he is illiterate and unable to add even a simple column of figures; you knew it when you hired him. So how can you call him responsible for it? He just isn't competent in these situations.'

And a final case:

'The law is that you are required to fill or cover wells on property which you own. The case against you is air tight. You are of age and sound mentally; you had access to that law—it was even written into your contract. You can read; you had plenty of time to fill the thing or have it covered. You did not. You are criminally responsible for that child's death.'

Each of these examples involves the assertion (or denial) that a man is a responsible agent. (Each of course involves more than that: for example, the assertion that he is or is not responsible *for* something.) And it is significant to note the variety of ways which the examples suggest for explicating the notion of responsible agency: one by reference to a man's mental state; another by reference to his physical abilities; a third by reference to his intellectual abilities; and a fourth by reference to several of the foregoing. In each case, it seems that only those abilities necessary for the performance or avoidance of the specific thing the man is being held responsible for are relevant to whether he is considered in that situation a responsible agent. 'He is a responsible agent' appears to mean 'He is able to behave in a specified way,' the way being specified by the situation—that is, the way being whatever necessary for a man's either avoiding or bringing about what he is to be held responsible for.

Indeed, it is difficult to imagine explicating responsible agent statements *apart* from rather specific situations, for though it is true that a man who cannot do anything (as one in a coma) might categorically be denied the status of responsible agent, it is also true that as soon as a man is asserted to be a responsible agent, it is proper (that is, not trivial) to ask under what circumstances, or in what situations, or to what extent he is a responsible agent. If he cannot read, but can speak and understand the spoken language, that limits the area of behavior in which he can be said to be a responsible agent; if he can lift one hundred pounds but not two hundred, that also defines limits for his liabilities; and so on.

So the question of whether or not a man is a responsible agent typically comes up in relation to the question of whether or not he is responsible *for* a specific thing, and the meaning of the 'responsible agent question,' at least in part, is, Did he have (or will he have) the ability to behave in such a way as to avoid or bring about the thing he is being held responsible for? That is, does he have the mental, physical, emotional, or intellectual ability to do or avoid doing things like this?

2 NECESSARY CONDITIONS OF RESPONSIBLE AGENCY

Sometimes, however, the responsible agent question comes out of a clear sky in the sense that the questioner has no intention of attributing or ascribing responsibility; he is just curious as to whether, and in what kinds of cases, he could attribute or ascribe responsibility to a man, if the occasion arose. In this sense, the responsible agent question is one about the general range of a man's abilities—whether he is of normal health and intelligence, whether he can read, and so on. One is tempted, here, to look for the minimum conditions of responsible agency in this general sense—to ask, 'What is the minimal range of ability necessary for a man ever, in any case, to be called a responsible agent?' But this is a most difficult question to answer, at least when one tries to list a set of necessary and sufficient conditions. The abilities required for responsible agency vary widely from case to case. On reflection, however, a few items do surface which could plausibly be called necessary requirements of responsible agency.

For example, it is clear that one cannot give a reasoned defense of calling someone a responsible agent if that person is not able, in *any* case, to think, give reasons for judgments, choose between alternatives, make decisions, and do what he decides to do in the everyday senses of these words. The attribution of these abilities to the man is, apparently, part of what we always mean by or presuppose by calling him a responsible agent. And there are fairly ordinary procedures for finding out whether these terms 'apply' to various men.

The other abilities we wish to call attention to by this very general use of 'responsible agent' may be summarized by saying that unless a man can understand the notion of responsibility—the notion of people's liability for sanction or interference—then he cannot properly be called a responsible agent. To explain: it should be clear that very often, when one says something like, 'Jones understands his responsibility, all right,' one does not mean merely that Jones knows *that* he is liable for sanction, or that he is liable for it in a specific case, or that whenever x occurs, he will be liable. One means more than that: namely that Jones knows *why* as well as *when* he is going to be held liable (whether he agrees with the justice of the reasons or not). A similar distinction for the business of understanding the whole notion of responsibility illumines one of the requirements we set for a man's being a responsible agent: we insist not only that the man know that the practice exists, but be able to understand at least some of the rationale behind it: for example, how, generally, the sanctioning is connected with the value of his conduct and with the fulfillment of his obligations. The fact that we do make such a requirement in practice helps to explain our refusal to hold infants and the lower animals responsible. For some pets (dogs, for example) appear to have an awareness of *when* they are likely to be punished; we do not, for that reason, call them responsible agents, or 'hold' them responsible for anything. They are not 'liable' for praise or blame, they just *are* praised or blamed as a training procedure. In order to be considered a responsible agent, then, the man must be able to be made aware of *why* it is that people are attributed or ascribed responsibility. Apart from specific situations, it is not clear that anything further can be said on the matter of what it means to say that a man is a responsible agent.

It should be emphasized, however, that in claiming that a man must have the ability to understand reasons for his being attributed or ascribed responsibility, two absurd claims are surely not being made: it is not being claimed that in order to be considered a responsible agent a man must *actually* understand the rationale behind the practice of attributing or ascribing responsibility; nor is it claimed that he must understand why responsibility is being attributed or ascribed to him in any arbitrarily selected case. Obviously, some people who are perfectly able to understand have never taken the trouble to do so; they are none the less responsible agents for this ignorance. Similarly, in some cases in which responsibility is actually attributed or ascribed to a man, there *is* no 'reason' for him to be able to understand because the attribution or ascription is capricious and unjustifiable. In those cases, when we say, 'He can't understand. . .' we do not mean to cast a shadow on his responsible agency. So it is quite important to explicitly retain the phrasing: 'the ability to understand *reasons for* attributions and ascriptions.'

It should be noted, further, that these requirements—the abilities we do in fact insist upon a man's having in order to go further with the practice of holding him responsible—are not merely arbitrary conventions. They are to be justified, in each case, in terms of the *point* of the whole practice of attributing and ascribing responsibilities. Procedurally, then, one secures the appropriate definition of 'responsible agent' by first valuing and then evaluating various options with respect to the rationale for the practice as a whole. These matters will be discussed in Chapter XV.

3 THE FREE WILL CONTROVERSY

It must be acknowledged that although arguments along the lines of the one just given have often been advanced, they have just as often failed to satisfy all those concerned with the question of responsible agency. It has often been 'felt' (and 'felt' is a more appropriate word than 'thought' here, for reasons to be made explicit in the next chapter) that there is some more fundamental issue, that the crux of the matter is not the range of a man's abilities, but rather the question of whether or not it makes any sense to say that he was able to do something other than he did, in fact, do. For, it is said, if a man could not have acted otherwise, he cannot be held morally responsible for the act. The argument, concisely put by Chisholm, goes as follows:

> 1. If a choice is one we could not have avoided making, then it is one for which we are not morally responsible.
> 2. If we make a choice under conditions such that, given those conditions, it is (causally but not logically) impossible for the choice not to be made, then the choice is one we could not have avoided making.
> 3. Every event occurs under conditions such that, given those conditions, it is (causally not not logically) impossible for that event not to occur.
> 4. The making of a choice is the occurrence of an event.
> 5. We are not morally responsible for any of our choices.[2]

The other horn of the dilemma is, of course, that if one were to have made a choice such that, given the conditions under which it was made, it was causally possible for some other choice to be made—i.e., given the deliberations leading to the conclusion to choose *x*, given the strong motivating feelings to choose *x*, given the attempt to choose *x*, still, if it were possible for him to wind up (against his will, desires, intuitions, and so on) choosing *y*, then his 'choice' is capricious, purely fortuitous. If his choices are fortuitous—if they just happen—then it is at least as difficult to maintain moral responsibility for them as for ones which are totally determined.

This, then, is the dilemma. If the causal hypothesis (premise 3 above) is true, then it follows that I was able to do only what I did in fact do; I am able to do only what I am in fact doing; I will be able to do only what I will in fact do. This seems to wipe out the legitimacy of the notion of a responsible agent—and thus a necessary presupposition of many attributive and deontic R-statements is lost. Yet if the indeterminist position holds for human behavior, one is in just as great a quandary—for then, no matter what one wants and tries to get, what he actually succeeds in doing will be only a matter of luck—not ultimately in his control, either.

4 THE NOTION OF AUTONOMY

The usual recourse at this point is either to attempt to find some 'middle ground' in which, although the agent's acts are causally effective, and effected by his intentions, desires, choices, deliberations, will, and so on, yet his initiation of these acts is autonomous with respect to the causal chain, or it is to argue that the fault is somehow in the insistence that the causal hypothesis (or perhaps indeterminism, for that matter) precludes the legitimacy of responsible agency claims. The latter attempt—an attempt to find fault with premise 1—is a tangled issue. A widely discussed attempt along these lines is the so-called 'defeasibility approach' developed by H. L. A. Hart and others.[3] Most of these issues will be avoided by the arguments of Chapter XV, and so there will be no need to get involved in them here. A brief comment on the former attempt—which might be called the 'autonomous agent' view—is appropriate, however.

Suppose that my 'agency' (my ability to consider alternatives, decide, and will acts) somehow operates on the causal chain (as an added determinant toward whatever happens) but is not itself in the causal chain. This language is loaded with dangerous metaphors, but the general idea is fairly clear—that the feature of our natures we call *will* or *agency* is not changed—not conditioned—by circumstances in the way supposed by a thorough psychological determinism (which would hold, I suppose, that our 'decision procedures' are all acquired and constantly conditioned by our experience). The will is supposed in some sense undetermined by antecedent conditions but (with other conditions) determinant of subsequent events. It has been argued that this notion of autonomy is enough to vitiate the force of the dilemma noted above.

But it should be clear that this argument is unsound. For the notion of an autonomous agent, however explicated, is in one form reducible to a causal hypothesis, and in another to a negation of that hypothesis. As follows:

Either the autonomous agency's operation is in principle describable by law-like statements or it is not. That is, the agency's operation is either 'law-like' or it is not. Suppose it is. In that case, the outcome (what is willed and eventually caused) may be said to be fixed, determined. For the input gotten by the agency is operated on in regular, inflexible, and determinate ways to yield an output which, given the input plus the operations on it, could not have been other than it was. And it makes no difference from the standpoint of the conflict with responsibility whether the 'principle' by which the will operates is absolutely free from influence by circumstance, absolutely 'unlearned,' or whether it merely sorts out and establishes priorities among alternatives. One still 'cannot help' what happens.

Now suppose the will's operation is *not* law-like. What can this be but simple and complete indeterminism? Surely this cannot be an avoidance of conflict with the requirements of responsibility. And it will not help at all, of course, to try to put these features together: to say that the agency's operation is in part law-like, or law-like but not such that every outcome is determined. One simply notes, in that case, that *in so far* as its operation is law-like, the results are determined, and *in so far* as its operation is not law-like, the results are indeterminate. In either case the results 'cannot be helped.' So the 'middle course'—or at least the autonomous agent version of it—does nothing to solve the free will problem.[4]

5 THE INTELLIGIBILITY OF DETERMINISM

In addition to the autonomous agent view and the attempts to get at premise (1) of Chisholm's argument, there is another group of attempts which probably should be mentioned. Their object is to show that determinism is itself an unintelligible, contradictory or perhaps just very unclear notion. Such attempts focus on matters like the fact that it is even difficult to formulate, discursively, an adequate understanding of the notion of a necessary connection. The best hope is, apparently, the analogy that can be drawn between causal connections and the form of a deductive argument in which one arrives at a conclusion which 'cannot be false,' given the truth of the premises and the validity of the argument. 'Insufficient cause' or 'acausality' may be understood on analogy with the form of certain inductive arguments (e.g., statistical explanations) in which the conclusion can be either true or false, given the truth of the premises and the 'validity' of the argument.

But the more serious difficulties arise with the attempt to formulate the 'causal hypothesis' (premise 3 in Chisholm's version), the so-called determinist thesis—and the corresponding thesis for indeterminism. There are difficulties here, for example, connected with the move from saying *some* events are determined to saying *all* of them are. It is not an easy task to get a statement of 'universal causation' which avoids a reduction to absurdity of the following kind: If it is true that I am able to do only what I do in fact do, then it is true *a fortiori* that I can believe only what I do believe, 'know' only what I do know, say what I do say. In deciding whether determinism is true, suppose I look at evidence and conclude that it is; you conclude that it is false (also on an assay of evidence). If the hypothesis is in fact true, neither of us could possibly have concluded otherwise than we did, for if we can only do what we do in fact do, then *a fortiori*, we can only 'conclude' what we do in fact conclude. How then can we distinguish true from false, belief from knowledge? The acceptance or rejection of the causal hypothesis on the basis of evidence depends on such a distinction if it is not to be a vain decision. Is this not paradoxical?[5]

Further, there has been some difficulty with the relation of determinism to predictability. Just one of the oddities in this regard, which some find to contain interesting suggestions for the free will issue, is that a rational human being is in a curiously self-defeating position with regard to predicting his own behavior. The point is that conclusions, convictions, and judgments are among those causal factors which determine our actions. While much of one's behavior is no doubt determined by factors he is not fully conscious of, and while it may even be the case that no action is ever fully determined by factors in consciousness, and while some conscious factors may have little or no determinate effect, still it is the case that the contents of consciousness are often part of the set of conditions which determine behavior. Now when I survey my circumstances and try to predict my behavior in some future situation, I add potential determinants to that situation. Every piece of evidence assembled with prediction in mind, every conclusion—however tentative—may, in effect, immediately invalidate itself in the following way: If I judge that factors *a*, *b*, *c* and *d* will cause me, under condition S, to do *x*, I have, in the very act of making that judgment, rendered the judgment suspect. For now I am *convinced* that I will be caused by *a*, *b*, *c* and *d* in S to do *x*, and that conviction may alter the situation. Some men, when told that an event is very likely to occur, resign themselves to it (thus, perhaps, making it more likely to occur), others increase their resistance and may succeed in preventing it. So I at least must reconsider my prediction and decide whether or not '*e*' (my original conclusion about what would happen), when added to the list of determinants *a*, *b*, *c* and *d*, will change the likelihood of my doing *x* in situation S. And, of course, once I decide whether or not it will change things, I have added another possible determinant to the list and must reassess again. This process is indefinite—an infinite regress whose consequence is that I can never come to a 'final' prediction of my behavior. Never, that is, unless I can by some method ascertain that no conviction about the course of my actions in situation S will affect whether or not I do *x* in S.[6]

Such apparent incoherencies in the notion of determinism lend support to the proposition that one need not be intimidated by the sort of attack on morality outlined in Chisholm's argument. Yet the way out of the dilemma is as yet no clearer than the determinist's thesis itself. A fresh start is needed.

NOTES

1 For discussions of responsibility using explications involving 'could have done otherwise,' the following are both typical and good-of-their-kind: relevant passages in G. E. Moore, *Ethics* (New York: Henry Holt, no date), and P. H. Nowell-Smith, *Ethics* (Penguin Books, 1954). Articles by J. L. Austin, 'Ifs and Cans,' in *Philosophical Papers* (Oxford: Clarendon Press, 1955), pp. 153–180; A. E. Kaufman, 'Moral Responsibility and the Use of "Could Have",' *Philosophy*, 37 (1962); Roderick Chisholm, 'Freedom and Action,' in K. Lehrer, *Freedom and Determinism* (New York: Random House, 1966), pp. 11–14.

The explication of responsible-agency using the pun 'response-ability' is neat, but a bit forced in some situations and loaded, perhaps, in favor of one horn of an unresolved dilemma in psychology. If the sheriff is held responsible for maintaining law and order, and one asks whether the sheriff is a responsible agent, he is asking, in part, about the sheriff's ability to *anticipate*. It is not clear whether anticipating can be adequately described in terms of response-abilities. At least there is some controversy in psychology on related issues. Until this can be resolved, it is best to avoid loading the present explication one way or the other.

2 Roderick Chisholm, 'Responsibility and Avoidability,' in Sidney Hook, ed., *Determinism and Freedom in the Age of Modern Science* (Collier Books, no date), p. 157.

3 For some of the literature on defeasibility, see: H. L. A. Hart, 'The Ascription of Responsibility and Rights,' *Proceedings of the Aristotelian Society*, 49 (1948–9), pp. 171–94; S. Stoljar, 'Ascriptive and Prescriptive Responsibility,' *Mind*, 68 (1959), pp. 350–60; P. T. Geach, 'Ascriptivism,' *Philosophical Review*, 69 (1960), pp. 221–5; George Pitcher, 'Hart on Action and Responsibility,' *Philosophical Review*, 69 (1960), pp. 226–35; Richard N. Bronaugh, 'Freedom as the Absense of an Excuse,' *Ethics*, 74 (1963–4), pp. 161–73.

4 For an interesting account of this sort of 'middle-ground' agency, see D. C. MacIntosh, 'Responsibility, Freedom, and Causality...' *Journal of Philosophy*, 37 (1940), pp. 42–51. But it still does not avoid the arguments against autonomy advanced here. See also: Nani Rankin, 'The Unmoved Agent and the Ground of Responsibility,' *Journal of Philosophy*, 64 (1967), pp. 403–8; and A. K. Stout, 'Free Will and Responsibility,' *Proceedings of the Aristotelian Society*, 37 (1936–7), pp. 213–30. One is hesitant to begin the involved task of relating what is characterized here by the definition of autonomous agent to Kant's arguments on these matters. Suffice it to say that it will be argued in the next chapter that a kind of therapeutic 'blindness' is best developed with regard to the causal hypothesis which some might argue bears a small resemblance, in limited circumstances, to the view that we must postulate the existence of an autonomous agency (or at least preserve our undeniable operational assumption of it) in order to reconcile ourselves to the notions of duty and responsibility at all. But the similarity of what follows to these other views is only a limited one, and not directed towards preserving one's 'sense of duty' at all, but at preserving one's inclination to respond in kind to benevolent and loving acts. Responses to injury can be handled adequately by other means. As for one's sense of duty, one is in fact 'freed' from the clutches of some sets of determining conditions by becoming aware of them, and that knowledge is as good for preserving one's sense of responsibility as postulating autonomy—at least if the postulation is consciously done. So the postulate is not needed for one's sense of duty or responsibility.

5 See, for a similar reduction, J. J. C. Smart, *Philosophy and Scientific Realism* (New York: Humanities Press, 1963), p. 126. Also, J. Bourke, 'Responsibility, Freedom, and Determinism,' *Philosophy*, 13 (1938), pp. 276–87.

6 See J. Canfield, 'Determinism, Free Will, and the Ace Predictor,' *Mind*, 70 (1961), pp. 412–16.

XV

AVOIDING THE FREE WILL ISSUE

With respect to justifying responsible agent statements, much of the tangled free will issue can be avoided. The task is to show how such avoidance can be safely accomplished. The arguments which follow are directed mainly at this avoidance task, undertaken not only as a discussion of justification procedures for responsible agency statements, but also as a step in the discussion of justification procedures for attributive and deontic R-statements.[1]

The core of the argument here is that the determinist's case against moral responsibility stands or falls not so much on matters of logic as on matters of rhetoric. The force of metaphors, and not the force of facts concerning causality, is what puts determinism in the service of moral scepticism. And the metaphors which have this result are not essential to the explication of the causal hypothesis for human behavior. The central concern, then, is with the choice of the various devices available for explicating the determinist's thesis.

It is helpful to begin by considering the point of holding people responsible. One asks for an answer to the question, 'Why (for what purpose or purposes) do we hold people responsible for their behavior?' By simple inspection of the practice one can distinguish at least three 'points' or purposes for attributions of responsibility, and we will begin with these: a compensatory one, a 'protective' one, and what might be called one of 'emotional satisfaction.' A consideration of these reasons (in some cases, mere motives) for holding people responsible helps to show just when and where the free will issue is relevant to the responsible agent question, and where and when it is not. The results for those interested in avoiding the controversy are promising, because the compensatory and protective purposes are quite compatible with the causal hypothesis—that is, the warrant for continuing the practice provided by each of these purposes is not at all affected by the causal hypothesis for human behavior. In fact, the protective purpose often presupposes the validity of the causal hypothesis. At least, it presupposes the validity of some hypothesis about human behavior which asserts a determinate relation between one's character, will, etc., and his acts. If there is some hypothesis which does this and which is intelligibly different from some combination of determinism and indeterminism, so much the better. No candidate so far has been convincing, however, for reasons advanced in the preceding section. The 'emotional satisfaction' aspect of the practice is in a more complex situation with regard to determinism, but not an impossible one from the standpoint of getting on with reasoned justifications.

I TWO ASPECTS NOT INTERFERED WITH BY THE DETERMINIST'S THESIS

COMPENSATION

We sometimes engage in the attempt to decide 'who is responsible' in the process of getting or making compensation. If a man steals from me, I may be interested in little more than getting my property back. It is not at all surprising to find that a man who has lost something very valuable is willing to bargain with the thieves, trading his power to have them punished for the return of the property. In such a situation, the issue of whether or not the thieves were 'metaphysically' free has no importance.[2] If the claim that x was unjustly taken can be established, then the return of x to me is right (other things being equal) regardless of the 'responsible agency' of the thieves. So the justice of holding them responsible in this compensatory sense is unaffected by the causal hypothesis for human behavior. The validity of the practice holds even if the causal hypothesis is refused by complete negation (i.e., the assertion of complete indeterminism).

PROTECTION AND EDUCATION

Sometimes, when one holds people responsible, he is interested in either re-inforcing, preventing, or changing their behavior. Some behavior is judged bad, some is judged good. If people do what is judged bad (and supposing it is bad enough to support interfering with them), one often holds them responsible—meaning that one considers taking steps to insure that they will not repeat their behavior. When their behavior is good, of course, one often wants to insure that it is repeated ('I hear you were responsible for that program; I just want to commend you on it. . . .'). The practice of holding people responsible, then, often functions simply as a means of protecting individuals and societies from more injury (by helping in a decision to put someone 'out of harm's way'). Or its function may be more than mere protection: e.g., to rehabilitate, or reform—to turn a disruptive element of society into a productive one. The point here is that when either educational or protective objectives are the purpose in holding people responsible, the acceptance or rejection of the causal hypothesis in no way destroys the justice of the practice. Indeed, any reasoned decision to use the practice either to educate or to protect will *include* the hypothesis that people's behavior is determined. Otherwise, the attempt to change behavior or influence results would be in vain. So far from vitiating this aspect of the practice, the causal hypothesis (or the mythical 'middle-ground' hypothesis) is necessary to it.

2 THE ASPECT WHICH IS INTERFERED WITH BY THE DETERMINIST'S THESIS

But there is another sort of situation in which we apparently find a use for holding our fellows responsible—even when all compensatory, protective, and educative purposes are ruled out. It is here that there is said to be difference between *its merely being justifiable* to punish a man for his conduct, and *his being morally responsible* for it. It is here that the determinist's argument comes to be of concern.

The situations I have in mind are the ones in which our exclusive focus is on giving people what they 'deserve'—situations in which, even if it were pointed out to us that punishment (or reward) serves no compensatory, protective, or educative purpose, we still would want to insist that the offender 'ought to be punished,' that he 'ought not to get off scot-free.' Why? one asks. 'Because he deserves it for doing what he did,' is the usual answer. Punishment (or reward) under such circumstances can only be called retributive, and when one asks what purpose it serves, he is met with either a reply in the spirit of Hegelian metaphysics (punishment as the negation of the negation, necessary to restore the 'balance' of things), or the infuriating repetition of the nature of retributive punishment: one simply punishes those who deserve it. Finis.[3]

But in practice, retribution obviously does serve some purpose for some people: namely, one of emotional satisfaction. Simple observation, simple recollection for most of us, will secure the claim that retribution has wide appeal for its emotional payoffs. It is precisely these emotional payoffs which are diminished significantly by one version of the causal hypothesis for human behavior—explaining nicely why the determinism controversy is of such insistent importance in retributive situations, why in these situations we must be so concerned with whether or not the one held responsible is (or was) a 'responsible agent.'

To explain: a real or imagined injury (unless it is so overwhelming as to immobilize) commonly generates a good deal of energy. We are 'galvanized' by it, as the pulp writers put it, whether to flee or to fight. Much scientific energy is presently being directed to studying related matters—the nature of aggression and its relation to territoriality, for example. These studies are far from complete. But it is sufficient for our purposes to note that in some situations, retaliation is psychologically appropriate for us in the sense that we 'get satisfaction' from it (a phrase often used, not surprisingly, in retribution-talk). The psychological mechanism involved is no doubt quite complex, and somewhat different from person to person. It probably has something to do with the simple release of energy punishment permits—whether that release is by direct physical action against the offender, or by some more vicarious route.

But the simple release of energy cannot be the whole story—else why would we worry about the nature of the agent punished? Why feel guilty about lashing out at 'those who can't help it'?

The most probable explanation—if the reader will pardon the oversimplification—is to the effect that we have internalized certain precepts about who is a fit subject for punishment and who is not, and that it just happens that the precepts enjoin us not to punish the helpless. (Such an injunction, of course, is not likely to be a mere happenstance. If one considers the cases in which punishment is likely to have some utility for behavior modification, one is forced to conclude that some modified version of the conventional legal usage of 'couldn't help it' comes close to identifying them.) In any case, when we become convinced, in a given case, that one who has stirred our retributive 'instincts' really 'couldn't help it,' our desire to punish is in conflict with our disposition to regard punishment in such situations as inappropriate. 'Satisfactions' are thus diminished.

The psychological mechanism for getting satisfaction from retribution may also have to do with empathy. It is plausible to suppose, for example, that one might vitiate one's rage by empathy with the culprit's suffering. This is only a crude hunch, but it does help to explain why punishing inanimate objects and non-responsible agents is ordinarily less satisfying than punishing responsible agents. One can only release energy in the case of those 'not sufficiently like us'; the sympathetic reaction is either non-existent—as with inanimate objects—or else leaves us, for some reason, with a sense of guilt rather than satisfaction, as in the case of punishing a mentally retarded child. But the punishment of 'responsible agents' who have injured us is somehow an effective way of venting our feelings. Now, obviously, when we are told that the causal hypothesis is correct, when we are told (convincingly) that every man is in reality like the helpless incompetent we get no satisfaction from punishing, that every man is such that he 'could not help doing what he did,' 'could not have done otherwise,' and so on, the retributive aspect of the practice is going to be (at least to a large extent) vitiated. It will be vitiated both with respect to punishment and reward. After all, what is the use (for venting one's feelings of gratitude) of rewarding someone who could not possibly have done other than benefit you?

What the determinist's argument does, then, is to claim that the 'couldn't help it' excuse applies to us all, in our every action. And to the degree that we accept the identity between the conventional 'couldn't help it' and the determinist's 'couldn't help it,' our emotional payoffs from retribution are interfered with in every case, and in the same way, as they are interfered with in conventional 'he couldn't help it' cases.

3 THE INTERFERENCE IS A RHETORICAL PROBLEM

Now we come to the crux of the matter. The interesting thing about all this is that the determinist must depend on a particularly persuasive way of stating his case in order to make us accept (or even think of accepting) the purported 'identity' between his 'couldn't help it' excuse and the conventional one. To wit: the usual way of stating the determinist's thesis in relation to questions of moral responsibility emphasizes what amounts to an analogy between humans and machines (or worse, between humans and puppets). The explication continually reminds us of what we *can not* do, *could not* have done. The modal 'could' operates here in a powerful way to suggest that we are helpless patients of circumstance. And that suggestion, colliding with our learned dispositions about punishment of the helpless, can trouble us deeply.

But suppose we express the determinist's position with the use of the modal 'would' in place of 'could.' Suppose we focus, as Bradley did,[4] on the fact that one does not raise uncomfortable fears by telling a man that, if he had a second chance, and if things were *exactly* the same again, he would do precisely the same thing he did the first time. A man is not uncomfortable with such a claim. Rather, he is likely to be reassured. For it is somehow comforting to think that one's decisions were, given the situation, the natural, the obvious, the inevitable choice. And that if things were exactly the same again, the same choice would be made. Yet what has been expressed here but the same sort of uncompromising determinism embodied in the causal hypothesis as stated at the outset: that every event occurs under conditions such that, given those conditions, nothing else could have occurred?

Take an example: Suppose a man has committed a crime. We are told of his miserable childhood, his existence in a subculture in which violating the law was commonplace—almost expected. We are told of his ignorance, his inability to imagine a way out of his misery, his alienation from every agency, every human contact trying to help him. And then we are told that, given those conditions, and the opportunity to commit the crime, he literally *could not have done otherwise.*

Then take the same case: review again the miserable childhood, the forces of environment, the ignorance of other alternatives, the alienation, and despair—and then draw the conclusion that, given those circumstances, *any similar person would behave as he did.* Given exactly those circumstances again, including the fact that he would not be 'morally reformed,' his choice *would be the same.*

I suggest that the first sort of conclusion, when elaborated, can trouble us seriously with respect to holding the man responsible, but that the second sort of conclusion does not. The first sort of conclusion, when elaborated with the metaphors it seems to beg for, conjures up a remorseless, ineluctable chain of events in which the man was at best a spectator to his own inevitable doom. But the second sort of conclusion, when elaborated, merely assures us that the act in question was not at all capricious, that it was, in fact, what we should have expected, and can continue to expect unless something is done. The elaboration of the first sort of conclusion interferes with the retributive aspect of the practice of holding people responsible. The second, while certainly not encouraging retribution, is neutral toward it.

This change in effect is not due merely to the change from 'could' to 'would' in the respective conclusions drawn. There are no doubt ways to emphasize the old helplessness images using 'would.' But the change *is* due to the images brought to mind: on the one hand, of humans as helpless patients; on the other, of a reassuring constancy and stability in human affairs. And it is crucial to note that the change in effect is brought about by a change in the rhetoric used, not by a change in the 'scientific adequacy' of the hypothesis expressed by the language. For to say that, given the same conditions, he would do the same thing again is to convey the sense of the causal hypothesis every bit as adequately as to say '. . . . he could not have done otherwise.'

The question now becomes, which 'version' of the causal hypothesis we ought to use—the one which interferes with the retributive aspect of the practice of holding people responsible (i.e., the 'puppetry version')? Or should we use the one which supports all aspects of the practice by emphasizing merely that human conduct does not occur in a haphazard, inexplicable way? The answer (matters of 'scientific accuracy' being equal) must obviously depend upon the values we can secure for the practice of holding people responsible. That is, in so far as the choice of rhetoric for determinism functions as a means to enhance or damage the practice (and in so far as its function as a means in the matter is the only thing to be considered) the choice to be recommended will depend upon the value of the end to be reached. While at first glance this seems to dictate against the version which enhances retribution (since retributive justice is distasteful to most of us—at least in the abstract), second thoughts reveal that it is not clear whether retribution ought to be eliminated—especially if what eliminates retributive punishment also interferes with our propensity to 'retributive beneficence' (pardon the ugly locution). It is not a pleasant thought to contemplate a world devoid of spontaneous return of love for love, kindness for kindness, gratitude for good deed. Yet it is fairly clear that the psychological mechanisms involved in retribution of one kind are very nearly like those involved in the other. (After all, if you couldn't help but have been kind to me—if you had no real choice in the matter—then my gratefulness seems somewhat out of place, just as does my vengefulness toward someone who has injured me through no fault of his own.)

Perhaps, then, and this is not suggested with total whimsy, we need to cultivate a sort of dispositional blindness to one version or the other of the causal hypothesis as the situation demands. In circumstances in which we have good reasons to want to avoid retributive conduct, attention to the 'helplessness' rhetoric can be useful. One laughs, of course, and with good reason, at the sudden and ludicrous vision of Gregory Peck, standing on the courthouse steps, getting the lynch mob to disperse by talking metaphysics. That, obviously, needs to stay in the realm of amusing fantasies. But the record of Clarence Darrow's summation for the defense in the Leopold-Loeb case is not a fantasy; and talk of metaphysics is precisely what Darrow did. He came down unremittingly on the 'helplessness' version of the causal hypothesis. And he won life-sentences for his clients. On the other hand, there are surely circumstances in which we want to preserve 'retributive' conduct—as in beneficence. The 'helplessness' rhetoric, if it has any effect at all here, is likely to have a damaging effect. So attention to another version of the hypothesis (if causality needs to be attended to at all in such situations) would be better.

4 THE ROUTE AROUND THE FREE WILL PROBLEM

On the matter of the responsible agency question, then, it is clear that two versions of the practice are either indifferent to the free will problem, or else positively require some sort of causal hypothesis (whether strict determinism or not). The third version, the retributivist or emotional satisfaction one, apparently requires for its effectiveness a dispositional 'blindness' alternating with awareness of one version of the causal hypothesis.

Each version of the practice will need justification—either as a right thing to do (under the circumstances), or as an obligation. The procedures for such justifications are the ones discussed in Chapters III to XI. The justification of the first two versions will not need to concern itself with the agency question at all—except to the degree that it can be shown that human conduct is *un*caused. But if the emotional satisfaction version of the practice is ever justifiable, steps will need to be justified for managing the rhetoric of the determinist's thesis.

It is with this last matter that some uneasiness may be felt, I suppose. For it might be objected that (putting aside the somewhat bizarre fantasies it generates) such therapeutic handling of the determinism issue is unphilosophic. For what it recommends, in effect, are consciously developed dispositions to ignore relevant bits of human knowledge of the world. But surely such an objection is unfounded. The selective ignorance recommended here is not to be equated with a callous suppression of an obvious truth. It is rather a manipulation of one version of an hypothesis which has a powerful effect on a part of moral life. It is not that one recommends a manipulation of the *extent* to which the causal hypothesis is explicitly applied to human behavior, only a manipulation of the way in which that application is made, and consequently what analogies (puppetry versus images of constancy) it calls to mind. What these procedures involve, then, is the cultivation of a dispositional refusal to call each and every available version of 'the truth' to mind in select cases. If this is a refusal to face 'the whole truth,' then so be it. We are used to making many such refusals just in the business of being courteous and tactful to others. We seem to have no trouble, in fact, justifying a great many 'screening' devices which enable people to be productive of what we judge to be moral good. The justification of a screening device in the case of the causal hypothesis—as long as it is simply a matter of choosing between two versions which are equally sound intellectually—should not present any unique problems at all.

It remains to ask about the relation of the agency question to the justification of deontic R-statements. The procedures, whatever the outcome, are the same used for attributive cases. One asks what the point of the practice is, then asks for a justification of each version of the practice found. The free will version of the agency question will either be relevant to the practice in the way it is relevant to the compensatory and protective points of making attributions, or it will be relevant in the way it is relevant to the emotional satisfaction aspect of the attributive practice. (At least, no other versions of its relevance are visible on this writer's horizon.) In either case, the issue may be dealt with, or bypassed, in just the way it is for justifications of attributive R-statements.

NOTES

1 For some other avoidance arguments, see A. E. Duncan-Jones, 'Freedom: An Illustrative Puzzle,' *Proceedings of the Aristotelian Society*, 39 (1938–9), pp. 99–120; Gardner Williams, 'Wrath, Responsibility and Progress in a Deterministic System,' *Journal of Philosophy*, 39 (1942), pp. 458–68; Arthur Pap, 'Determinism and Moral Responsibility,' *Journal of Philosophy*, 43 (1946), pp. 318–27; and O. C. Jensen, 'Responsibility, Freedom, and Punishment,' *Mind*, 75 (1966), pp. 224–38.

2 However, the issue of whether or not the act is to be called a theft might be settled, in part, in that way.

3 For more or less adequate coverage of the issues raised in the extensive literature on retributive punishment, see: J. Mabbott, 'Punishment,' *Mind*, 48 (1939), pp. 152–67; H. J. McClosky, 'The Complexity of the Concepts of Punishment,' *Philosophy*, 37 (1962), pp. 307–25; S. I. Benn, 'An Approach to the Problems of Punishment,' *Philosophy*, 33 (1958), pp. 325–41; A. Flew, ' "The Justification of Punishment",' *Philosophy*, 29 (1954), pp. 291–307;

Kurt Baier, 'Is Punishment Retributive?' *Analysis*, 16 (1955), pp. 25–33; W. G. McClagan, 'Punishment and Retribution,' *Philosophy*, 14 (1939), p. 281. Ted Honderich's discussion of 'reactive attitudes' in *Punishment: The Supposed Justifications* (London: Hutchinson, 1969), pp. 115–23, is illuminating.

It must also be noted that some writers have argued that a man has a *right* to punishment—that to be 'reformed' is to be handled like an object while to be punished (under certain conditions) is to be afforded some semblance of human dignity. See, in this regard, Herbert Morris, 'Persons and Punishment,' *The Monist*, 52 (1968), pp. 475–501.

The procedures here go at the question of a man's rights by way of justifiable institutions in just the way the question of one's duties is dealt with. If one can establish the justice of a man's right to punishment in this way, then no reasoned and countervailing protest could be raised. Indeed, there is a good deal of cogency in attempts to do this—as well as in the reminder that occasionally a man (from whom we have nothing to fear in the future) needs to be punished to help him deal with his guilt over his wrongdoing. It is uncertain, however, that one can find a corresponding version of such points in the case of attributing responsibility for 'right-doing.' Do we want to say a man has a right to praise? To compensation, of course. But praise? And is there some unfortunate result of righteousness for which one needs the help or the praise of others? Perhaps, but not obviously. One can understand that people need approval for their good efforts in order to keep up interest in doing them, but when that is not at stake (i.e., when we have 'nothing to fear' in this regard, corresponding to the case of the guilty wrongdoer), it is hard to spot a 'need' for praise in a sense corresponding to the guilty man's need.

Here one again feels the need for co-ordination with deontological and agent morality approaches. For one suspects that it is the resistance to the pre-emption of one's *agency* which leads us to resent the idea of 'correction' more than that of 'punishment.' (Both are violations of one's person, of course, but correction is doubly so.) The sort of benevolent (unneeded) praise just mentioned is best dealt with in terms of moral character.

4 F. H. Bradley, 'The Vulgar Notion of Responsibility in Connection with the Theories of Free Will and Necessity,' in his *Ethical Studies* (reissue of the second edn, Oxford: Oxford University Press, 1962).

XVI

JUSTIFYING ATTRIBUTIONS OF RESPONSIBILITY

The analysis of attributions concluded that the central and necessary feature of the act was the assertion of someone's liability for sanction or other interference. Consequently, there are at least two items to justify for any attribution: (a) that what happened to occasion the attribution is the sort of thing for which someone may properly be held liable for sanction or interference; and (b) that the particular person(s) picked out by the attribution are ones properly held liable in the case at hand. In addition there are, of course, at least two other items which very often must be considered: the question of cause and the question of responsible agency. These have been duly noted, but to repeat: the procedures for establishing the former are not at all special to morals; the procedures for dealing with the latter have been outlined in Chapters XIV and XV.

1 THE OCCASION FOR THE ATTRIBUTION

The justification of (a) above hinges on two points: one is whether what happened has a clear value (i.e., whether it is either good or bad rather than neutral or so mixed as to allow no judgment as to which is preponderant), or in the case of conduct, whether it was right or wrong; the other is whether what happened is appropriately dealt with by sanctioning or otherwise interfering with person(s), and if so, which person(s). The procedures for handling the first of these two points have been outlined in previous discussion. It may simply be remarked here that if an event cannot be said to be either good or bad, or a piece of conduct either right or wrong, it seems clear that sanction for it (or interference of any sort) is out of the question if for no other reason than that the lack of such judgments means the lack of the usual criterion for guiding the decision on what sanction (i.e., punishment or reward) or interference might be appropriate. Sanctioning is normally guided by the supposition that the interference is different, depending on the value of the conduct, and depending on what sort of interference is good or right for conduct of the given value. (This wordy criterion is meant to encompass both utilitarian and retributivist positions.) But suppose something happens in a plant (the statistics on production shift in some way) which no one is able to find a value for one way or the other. One lacks the usual criterion for deciding, then, whether to blame or praise, punish or reward, and the notion of compensatory action (as distinct from action simply taken to return things to a status not demonstrably different in value from the new one) has no warrant. (Of course if it is argued that the plant manager was wrong to allow a change to such a confusing state of affairs, then there is a criterion: the wrongness of his behavior.)

The procedures relevant to handling the second point are those outlined for the justification of an institution. These procedures are applied to the question of whether the attribution practice can ever be justified at all for a case such as is at stake. Then they are applied to the question of the range of persons 'connected' to the case which attribution may, *ceteris paribus*, be extended: that is, do they properly extend to the sons of the sinning father? To the teachers and parents of the wrongdoer? To the man who incited the riot as well as the rioters?

This question of the *extent* of the liability is crucial, for quarrels on this matter occupy a good deal of the time spent in both law making and adjudication. It is important to see that the grounding of judgments about extent of liability, as well as those about the inclusion of various types of cases in the attribution practice *per se*, are achieved (if at all) by way of justifying the use of attributions, in various types of cases, as an institution, an institution defining not only what to do (attribute) and when, but defining also *whom* to do it to (the extent of the liability).

Even though the outlines of the procedure discussed in the

preceding paragraph are clear enough, it might be appropriate to indicate briefly the 'substance' those procedures are likely to be put to work on. The general structure of the considerations is likely to be given by the aspects of the attribution practice outlined in Chapter XV. The specific question to be asked in that case is, 'In a situation such as this, is there any way that liability for sanction can be made to serve the compensatory (or protective or emotional satisfaction) purpose of the practice?' This calls for an answer in terms of the axiological scheme, and the answer must be that if the purposes can be justified as good (and the practice in those aspects judged to be right) and liability for a specific course of sanctions or interference (in a given situation) can be shown to help achieve the purpose(s) involved in the practice, then the attribution practice has surely been shown to be 'good for' handling the case, *ceteris paribus*. Other valuations of it can of course be made without reference to the point of the practice. The values for the practice are then evaluated to determine if it is the right thing to do, and whether one ought to do it in various types of cases. As always, such procedures are open to dispute by way of their *ceteris paribus* clauses.

The next item is, of course, to ask what sort of reasons to the contrary are likely to fill the *ceteris paribus* blanks in such arguments. A general characterization of the range of 'reasons against' is a difficult matter, but the following remarks may be helpful. If one can show that, even though the purposes of the practice are legitimate, and the practice helps to achieve the purposes, there is a 'better' way to do the same thing (or a better thing to do altogether), this would obviously be a countervailing reason. It is at this point that the presupposition of responsible agency contained in some attributions comes up. In certain cases the legitimacy of the practice (i.e., its very effectiveness in carrying out its purpose) is contingent on whether or not the one held responsible is a responsible agent (not in the 'free will' sense, but the sense of having certain specific abilities). So an objection to the procedure—in effect a 'reason to the contrary'—may be that in a particular case the definition of 'responsible agent' made necessary by the nature of the case is such that no normally formed human can be expected to meet it—e.g., the ability to run a mile in one minute. Evidence for this contention will often have to be drawn from empirical psychology and physiology, but its effect will be to show that in some cases, because sanctioning can have no conceivably desirable effect ('Punish all you want, Ryan won't be able to do a mile a minute'), holding a man liable for sanction is pointless.[1]

2 PERSONS NAMED IN THE ATTRIBUTION

Suppose, now, that the legitimacy of holding some sorts of people liable for sanction or interference in type-of-circumstance S has been established. We come now to the question of whether this or that particular person can justifiably be held liable in such a circumstance. Once that is determined, the issue moves to the justification of actually imposing the sanction.

The first matter to arise is inevitably the one of responsible agency. But we have seen that this is not always relevant to justifying attributions. So the task is to find out if the attribution presupposes responsible agency, and if so, what counts as responsible agency in the case under discussion. Does the agent need to be able to read? Or will a speaking use of the language do? (Recall that the free will version of the responsible agent question is being avoided.) The procedures for establishing agency here are not at all special to moral philosophy, for there is no valuation involved. At least, if there are questions of value ('He has to be a good swimmer'), they are of the sort one is accustomed to handling in almost any field of inquiry: e.g., by restating: 'He has to be able to swim 100 yards in three minutes.' If agency is presupposed by the attribution, to have shown its presence is merely to have opened the way for argument on the legitimacy of the attribution. Of course, if agency is not presupposed, the way is already clear. One may, of course, ask again on what grounds agency is justifiably presupposed by some attributions and not by others.

Once the agency question is settled, the most economical procedure is to ask (1) whether the person(s) named in the attribution had a (deontic) responsibility for what happened; (2) whether he (they) failed to meet that responsibility; and (3) whether there are any reasons to be found for excusing him from liability (not excusing him merely from sanction, now, but even from liability for it).

The reason this procedure is economical is that it organizes the relevant concerns nicely. Many attributions (one wants to say most, but a count would be difficult to make) can be seen as the logical, though not always psychological, outgrowth of corresponding deontic responsibilities. If it can be said that I have a responsibility to do *x*, and I fail to do it, then the attribution for not doing *x* follows (*ceteris paribus*). If responsibility is attributed to me for *y*, often it can be said that I have (or had) a responsibility for seeing to it that non-*y*.

By first establishing the deontic responsibility and then carefully noting its requirements, much argumentation can be shortcutted (e.g., 'But I didn't do anything.' 'Sorry, but your responsibility is not only to refrain from doing *x*, but to take steps to see that *x* doesn't just "happen all by itself." ' Or, 'But I didn't do anything.' 'Sorry, but your responsibility extends to the actions of your staff as well.'). The range of possible excuses ruled out (or in) by an examination of the terms of the deontic responsibility is significant. Not only matters of causation (and haggling over proximate or secondary cause) may often be handled here, but also questions about the relevance of excuses concerning intent (or lack of it), and some of the 'I couldn't help it' excuses. Often, however, these excuses are more relevant to the question of whether to sanction than to the question of whether one is liable for it (see below). In any case, if there is a corresponding deontic responsibility, it usually has to be justified in the course of arguments over the attributions, anyway, so it is not likely that this procedure takes on the proof of more than would normally be required.

Once the corresponding deontic responsibility is established (step (1) above), the next step is to establish that the person(s) to whom the responsibility is attributed failed to meet that responsibility. Surely if a man has not in any sense failed to meet his responsibilities, we cannot from the mere fact that 'something happened' hold him (as opposed to someone else) responsible. That is, we 'cannot' in the sense that from those items above we do not yet have a reason—justification—for doing so. There may of course be reasons unconnected to his having had a deontic responsibility which justify holding him liable; this possibility will be discussed below. The point here is that to justify holding a particular man responsible rather than another requires the exhibition of some kind of connection between that man and the event. It so happens the most general way to state the most frequent type of connection is to say that he both had and failed to meet (or had and met) a deontic responsibility for the event in question.

There seems to be slim chance of coming up with a useful set of necessary and sufficient conditions for all attributions of responsibility—even one which is simply descriptive of current practice. It would be nice if one could do this, for then he would have an organized frame for inquiring into whether current practice in these matters is justifiable. If it is—and if it is expressible by a set of necessary and sufficient conditions—the casuistical business of adjudicating excuse attempts would be much simplified.

But no such set of necessary and sufficient conditions seems obtainable for current practice. Consider: there are cases (e.g., forgetting) in which we attribute responsibilities in the apparent absence of intent to do wrong, the absence of anything that could reasonably be called an act, voluntary or not, and the absence of foresight ('But I couldn't have foreseen my forgetting' will not do as an excuse). If foresight, intent, and action are not necessary for attributions, one hardly knows where to turn next. 'Cause' is obviously not a candidate, for responsibility is often attributed for 'what happened' (liability for an injury which occurred on one's property) where there is no question of one's having 'caused' anything, even by negligent behavior.

These four items are the most likely candidates for a necessary condition, along, perhaps, with one other: the absence of compulsion (psychological or physical), the absence of overpowering coercion, and the presence of some option to the conduct actually done. Leaving aside the matter of whether 'free will' may be defined along these defeasibility lines, the set of considerations noted here obviously indicates an important source of excuse attempts and is an obviously necessary condition for the justifiability of *some* attributions.

But it, like the others, will not do as a necessary condition of all attributions, for there are cases (e.g., in which one first intends to do what he is later compelled to do) which render subsequent compulsion, coercion, or lack of option irrelevant. A man who fully intends to commit a crime and is then forced, against his will, to take drugs, undergo hypnosis, etc., and then to commit the crime is not, at least not clearly, excusable from responsibility for the deed when it is once accomplished. It will not do to say responsibility is attributed to him only for his intent (his subsequent 'act' being out of his control). For (as long as one is asking, now, about what people do, not what can justifiably be done) it seems clear that some people would be inclined to regard the whole intervening drug episode morally irrelevant. If this is so, then the existence of compulsion, etc., is simply not an excuse in some cases, and the absence of it is not a necessary condition. Of course the point is more easily established by reference to situations (e.g., some monetary compensation cases) in which the question of responsible agency does not even arise.

As for sets of sufficient conditions, one can find cases in which intent alone is sufficient, and cases in which it is not;[2] cases in which intent is sufficient when combined with any selection of the following: foresight; causing a wrong; just acting abortively (attempted crimes); and the last mentioned set of considerations relating to free will (may one call their presence volition temporarily?).[3] But there are also cases for which each combination is not sufficient.[4] There seem to be no cases in which volition, foresight, cause, or acting abortively—each taken alone—are sufficient. But volition and foresight are occasionally sufficient (cases of negligence), and any combination of three can be (but is not in all cases) sufficient.

Whether this abrupt review raises more questions than it answers or not (moral psychologists who have devoted books to 'intent' and 'action' are no doubt quite displeased by now), at least it shows the likelihood that one will simply have to define the range of excusing conditions case by case—perhaps by defining the corresponding deontic responsibilities in a way that specifies the range.[5] One would insist in each case, that the deontic responsibility not only define when, where, and who is liable, but under what (personal) circumstances a man is liable. This last amounts, in part, to specifying the abilities necessary for responsible agency in each case. And it would then go on to specify what will and will not count as an excuse, given that the man was a responsible agent in the specified way.

The final step, then, is a version of the familiar *ceteris paribus* investigation—in this case in the form of asking whether there are any factors present which, as defined by the deontic responsibility, excuse the man from liability, i.e., which countervail the justification of liability. This step presents no peculiar procedural problems, though one may find it convenient to treat some frequently heard excuses (compulsion, insanity pleas, lack of intent to do wrong, superfluousness of the sanction) as excuses from sanctioning itself rather than as excuses from liability.

What about attributions which cannot be dealt with as 'outgrowths' of a failure to meet a deontic responsibility? Might there not be cases in which we want to hold a man responsible even though he cannot be said to have had an obligation in the matter? Perhaps so. If so, the procedure must be (1) to establish some 'connection' between the man to whom responsibility is attributed and the event which justifies picking him out for special treatment as a consequence of the event (perhaps, here, a causal connection); and (2) to make sure there are no reasons which excuse him from liability. One cannot be more specific than that until he is presented with a case, and without the definition of a deontic responsibility to guide matters, the procedure is likely to be 'loose' or 'open-textured' in the extreme. The question of whether the punishment of the innocent may ever be justified will be dealt with in Chapter XVII.

NOTES

1 See H. L. A. Hart, *Punishment and Responsibility* (Oxford: Oxford University Press, 1968), Chapter II, for an important argument on the reasonability (in terms of maximizing each individual's control of his destiny) of exempting certain sorts of people, as well as certain sorts of acts, from attributions of responsibility.

2 Sometimes we find vicious intentions blameworthy (as when an adult 'fully intends' to cheat you); but at other times we do not take them seriously enough to blame anyone (as when someone who is mentally ill is spiteful).

3 *Intent with foresight*: same sort of examples as for intent alone. *Intent with causing a wrong* and *acting abortively* are common causes. *Intent and volition* can best be illustrated by imagining how we might react to the 'mental patient's' spitefulness if it turned out that he wasn't a patient at all, but a visitor to the ward.

4 Consider the case of a pathetic eccentric, trying to control his M.P.'s vote on the Witchcraft Bill with magic. He certainly has volition and intent. But even if he tried to kill the M.P. with magic, he would not be charged with an attempted crime.

5 For further studies on related issues, see Part I of H. L. A. Hart and A. M.

Honoré, *Causation in the Law* (Oxford: Clarendon Press, 1959); Frederick Will, 'Intention, Error, and Responsibility,' *Journal of Philosophy*, 61 (1964), pp. 171–9; Bloomenfeld and Dworkin, 'Punishment for Intentions,' *Mind*, 75 (1966), pp. 396–404; and J. Bensen, 'Characterization of Actions and the Virtuous Agent,' *Proceedings of the Aristotelian Society*, 63 (1962–3), pp. 251–66. Also, of course, J. L. Austin's 'A Plea for Excuses' in his *Philosophical Papers* (Oxford: Clarendon Press, 1955), and the extensive literature on intention, free action, etc., in recent work in philosophy of mind—e.g., G. E. M. Anscombe, *Intention* (Oxford: Basil Blackwell, 1958).

XVII

JUSTIFYING SANCTIONS

The next question concerns procedures for justifying the actual imposition of sanctions or other interference. The discussion will often sound as though it is simply punishment, or perhaps blame, that is at issue, but here as in the preceding sections it should be understood that the punitive sanctions or interferences referred to cover a much wider range than is usually conveyed by 'punishment' or 'blame' (e.g., all sorts of educative devices designed to alter an offender's behavior, many of which may be quite pleasant); further, all that is said about justifying the punitive sanctions applies, with insignificant modifications, to the 'rewarding' sanctions.

Two forms of the question about justifying sanctions will be considered: one which assumes that one to whom responsibility has been attributed has been shown to be liable for the sanction, and asks simply whether it is justifiable to go ahead and apply the sanction; the other which does not presuppose that the man is liable (either he has been shown not to be, or it is not known whether he is), and asks whether sanctioning him might be justified anyway (e.g., punishing him 'as an example').[1]

1 JUSTIFYING SANCTIONS FOR ADMITTED LIABILITY

When the man has been shown to be liable for sanction, two things must be shown to justify proceeding: (1) that the specific sanctioning proposed would probably serve its purpose; and (2) that there are no reasons which warrant excusing him from sanction. The two matters are by no means always distinct; one may find an excuse from the fact that the sanctioning is not likely to work. But it is convenient to make the division for discussion.

THE PURPOSE OF SANCTIONING

When the purpose to be served is the educative-protective one, the first of these points is a notoriously difficult matter in the social sciences: what techniques are effective in guaranteeing a change in a given man's behavior? Punishment (so-called 'negative reinforcement') sometimes has the effect of entrenching the offender in his conduct rather than reforming him. On the other hand, it turns out that some of the mentally ill (whom men of good will have worked so long and hard to exempt from punishment) are helped most effectively by a therapy which includes fairly crude versions of punishment-reward training. Whether a given man will be 'reformed' by a given sanction (or whether a given prison will hold him if a mere custodial sanction is proposed) is, then, an empirical question and one usually requiring expertise. When other purposes for sanctioning are considered (i.e., compensatory or emotional satisfaction purposes) the justification is likewise empirical, though often (especially in cases involving monetary or 'goods' compensation) much more straightforward.[2]

EXCUSING CONDITIONS

Once it has been established that sanctions would work, one looks again for reasons which would warrant excusing the man from sanction. Such excuses divide roughly into the following types: 'But there is a better way'; 'But it is unnecessary (or pointless or unutilitarian) to do anything about this case'; 'But he couldn't help it'; 'But he couldn't have known'; 'But he didn't mean to do it.' A great deal of writing has been done about such excuses, most of it description—attempts to lay out when each of the excuses is accepted in practice and when not. But what is wanted is an indication of the connection, if any, between the various types of excuse and the practice of holding people responsible—an indication of why one might judge them to be adequate excuses, and thus how they fit into the justification procedures.

'But there is a better way' calls into question the whole enterprise of sanctioning in this case (saying that the same purpose, or a better one, can be achieved in another way which, viewed alongside sanctions, must be valued as the better, i.e., the right, course of action); or it claims that, although the proposed sanction would work, there is a better method of sanctioning or interference to be found. (To claim that the proposed sanction would not work would be to dispute step (1) directly; not, strictly, to make an excuse in the way the questions have been divided here.) The claim each of these excuses makes on action is just the claim that the original justification of the practice of holding people responsible has made: it was shown that the practice was good, bad in no countervailing way, thus right, and thus what one ought to do. Here it is claimed (in the first version of the excuse) that another course of action has been found which, at least in the case at hand, is not only good, but on balance better than the final part of holding the man responsible (i.e., sanctioning him). It is, then, the more appropriate thing to do (closer to the right thing), and ought to be done. It is asserted that one has simply failed in the evaluation of the possible courses of action in this case—that although the proposed one may be good, it is not the best and therefore not *the* (though possibly *a*) right one. In the second version of the excuse it is likewise argued that one has failed in the evaluation of the possible sanctions—that although the one proposed is good (and some sanction ought to be imposed) it is not the best and therefore not the right one. It is asserted that there is a better—i.e., more appropriate—one.

'But it is not right to do *anything* in this case.' This excuse also has two forms: one a claim that the sanctioning is unnecessary; the other that it is for some reason to be disvalued. Both forms claim an error in evaluation. In the one case it is claimed that (for example) the view that sanctioning would work on the man may be correct, the view that holding men of his sort responsible for things of this sort may be correct, but that the actual sanctioning in his case is superfluous because he has already reformed, made compensation, or whatever (i.e., fulfilled the point of the procedure). It is claimed that the evaluation simply was not carried far enough. This excuse is often heard in cases involving the humiliation of otherwise good men who have erred: 'He has suffered enough.'

The second version of the excuse is a claim that, although the sanctioning may be in fact the best conceivable way of dealing with the offender, still it has disadvantages for others which outweigh its good (it is too costly, or too much trouble). Although this claim is often heard and seems unobjectionable procedurally, one must take care, for it is very close kin to arguments in favor of punishing the innocent for some 'greater good.' And though most occasionally find it easy to accept letting the guilty go if punishing is too bothersome, few ever find it easy to accept punishing the innocent.[3] The question will be examined more fully below.

'But he couldn't help it.' Much consideration has already been given to this excuse. The determinist's version of it has, after some detailed argument, been turned aside as far as justification procedures go. It has been pointed out that its initial relevance is usually to be determined in the process of deciding whether the man is liable for sanction—more specifically in connection with examining the nature of the deontic responsibility violated by 'what happened,' for some deontic responsibilities recognise no excuse of this sort, others are mute on the question. When they are mute, or when for some other reason the consideration of the issues of compulsion versus 'volition' do not come up in the justification of liability, they certainly must be considered at this point. Here the relevance of the excuse is to the potential effectiveness of the sanction. It claims that something important has been omitted from consideration in that, although it may be true that sanctions applied to this 'sort' of offender for this sort of wrong are effective, they are not effective when the man 'could not help' (i.e., was compelled) to do what he did. In those cases, it is argued, the sanctioning only makes him bitter, not better. The adjudication of this claim is again a matter involving empirical psychology. For it may be (as in the case of forgetfulness, where it is often claimed that one 'couldn't help it') that the sanction still will change some habits of behavior which will prevent the recurrence of the offense.

'But he couldn't have known.' This attacks the legitimacy of claiming that a man violated his corresponding deontic responsibility. But it functions against the effectiveness of sanctions in just the way the 'I couldn't help it' excuse does. It serves to invalidate any educative point to the practice of sanctioning, for, unless through sanctions the offender learns to pay more attention to his tasks and thus 'to see farther ahead,' one can hardly use the sanctions as a means for rehabilitation. So unless sanctions have the effect of improving foresight, the 'I couldn't have known' excuse can be a powerful one. It probably can vitiate the emotional satisfaction aspect also (as does the 'I couldn't help it' excuse) for the same reasons that the determinist view does. But it does not take effect against the compensatory and purely protective (getting him out of harm's way) purposes of the practice. For these sanctions 'work' (accomplish their purpose) regardless of the bitterness, for example, of the offender.

'But he didn't mean to do it.' The legitimacy of this excuse runs exactly parallel to that of the 'couldn't have known' excuse.

Presupposing the justification of a man's liability, then, the actual sanctioning is justified if: (1) it is established that the sanctioning would work; and (2) it is established that no excuses exist to block the justification. It remains to consider whether or not a man whose liability for sanction has not been established may justifiably be sanctioned.

2 JUSTIFYING SANCTIONS FOR ADMITTED NON-LIABILITY

One of the objections to the attempt to deal with questions of punishment (and, one supposes, the question of responsibility in general) solely on a utilitarian basis is the suspicion that it might happen that 'the greatest good for the greatest number' requires a scapegoat—the punishment of an innocent man for something he did not do (or is no more responsible for than anyone else). This situation, it is argued, is simply intolerable. It does no good to argue that the concept of punishment presupposes guilt on the part of the one sanctioned (otherwise it is not punishment at all, but gratuitous injury). This does no good, critics say, because it merely forces a restatement of the problem: now what is monstrous is that the procedure may permit the justification of gratuitous injury. Does not the same objection apply to the present procedures—even though they are not strictly (i.e., solely) utilitarian?

The reply is that the objection is relevant to the present analysis in the sense that injury to the innocent cannot be ruled out on procedural grounds alone. This in itself is not an objectionable feature, surely, but the objector's suspicion may be that these procedures seem likely to justify injuries to the innocent helter skelter. Such suspicion is, however, unfounded, and can be blunted by noting the following: first, just in terms of the axiological scheme alone, any affective valuation against injuring the innocent will enter into the balance, just as other valuations of a more cold-blooded sort do. The sadist's valuations do too, of course. But second, one must not forget that the valuations of the innocent man will also enter into the balance on equal feet with all others. And the principles from presumptive value criteria governing the 'weighing' of one valuation against another combine with this fact to favor the valuations of the innocent man over any of the least bit 'whimsical' valuations of his would-be prosecutors. The criteria favor, it will be recalled, permanence, pervasiveness, fecundity, and so on, all features on which the innocent man's valuations are likely to be based. Similar considerations apply to the use of the presumptive deontic criteria, and all this, when added to the considerations to be advanced below concerning agent morality, combines to reassure one that 'punishment of the innocent' is not likely to be justifiable.

Finally, of course, in order to justify sanctioning any innocent man *x* (as opposed to innocents *y* or *z*) one needs a reason which picks him out from the others. To say that one can establish that it would be justifiable to sanction *someone* is not enough to establish the justifiability of sanctioning anyone in particular.[4] One needs to establish the more difficult case that it would be justifiable to sanction any one of a specified class, that *x* is a member of that class, and that there are no excusing conditions. When this is combined with the probability of the use of these procedures to establish a substantive moral principle prohibiting punishment of the innocent, much of the urgency of this objection is lost. However it must be admitted again that the justifiability of injury to the innocent cannot be ruled out on procedural grounds alone.[5]

NOTES

1 I am indebted to F. E. Sparshott for pointing out to me that no discussion of the justification of sanctioning can be complete without some consideration of the 'legitimacy' of the agency (person or group or organization) which does the sanctioning. We typically can offer good reasons for not wanting mobs to impose sanctions—even on the 'guiltiest' of criminals. Questions of this sort are straightforwardly amenable to the justification procedures presented here, however.

2 On the point of sanctioning serving its purpose, see D. Braybrooke, 'Professor Stevenson, Voltaire, and the Case of Admiral Byng,' *Journal of Philosophy*, 53 (1956), pp. 787–96; L. Kenner, 'On Blaming,' *Mind*, 76 (1967), pp. 238–49; O. S. Walker, 'Why Should Irresponsible Offenders be Excused?' *Journal of Philosophy*, 66 (1969), pp. 279–90; and two articles by Vinit Haksar, 'The Responsibility of Mental Defectives,' *Philosophy*, 38 (1963), pp. 61–8, and 'Aristotle and the Punishment of Psychopaths,' *Philosophy*, 39 (1964), pp. 323–40. Also, again, Hart, Chapter II, *Punishment and Responsibility* (Oxford: Oxford University Press, 1968).

3 The 'It is best not to do anything' excuse is a delicate matter in at least two additional ways: the wrongdoer may 'need' the punishment as a means of handling his guilt; and he may have a right to it, as some have agreed. One can sometimes ignore these matters in a benevolent zeal to excuse.

4 Recall one of Yossarian's problems in Joseph Heller's *Catch-22*. Yossarian understands, he says, why the Allies must win the war, and he understands that in order to win the war, some men are going to have to die. What he doesn't understand, he says, is how it follows that *he* has to die. Precisely.

5 See also R. S. Downie, 'Social Roles and Moral Responsibility,' *Philosophy*, 39 (1964), pp. 29–36; and Richard Brandt, 'A Utilitarian Theory of Excuses,' *Philosophical Review*, 78 (1969), pp. 337–61.

XVIII

JUSTIFYING ASCRIPTIONS OF RESPONSIBILITY

The next subject is the justification of deontic responsibilities. Most of the procedures applicable here were outlined in the section concerned with the justification of obligations, for 'obligation' was given very wide interpretation which included both duty and deontic responsibility. One considers, then, the deontic R-statement as derivable from 'institutional facts' and the definition of the institution is left to the ordinary valuative and evaluative procedures. Several elaborative remarks are called for, however.

1 ACCEPTING RESPONSIBILITY

First, it should be noted that a difference of procedure might be supposed to exist between the justification of a responsibility that a man can be said to have accepted, and the justification of those where no acceptance was present. Responsibilities are often occasioned by a man's position, role, occupation, or commitments. When such things can be said to have been undertaken by the individual, and can be shown to entail the deontic responsibility in question, it might be assumed that one can reasonably operate from a presumption (subject to contradiction or excuse) that the man has the responsibility. This is a convenience in that it ordinarily shortens the procedure: justifications of the 'institutions' involved are not usually demanded by the man who has accepted responsibility. He usually agrees to their justifiability. But it must be stressed that such conveniences have dangers. To rule out consideration of part of the justification procedure merely because no questions about it are raised by the parties involved is to take no guard against the well-known short-sightedness of the 'interested parties.' Thus it seems wisest to ignore the acceptance factor and begin from scratch. Consideration of whether or not the man accepted the responsibility may be brought in during the search for excuses.

2 DESCRIBING THE RESPONSIBILITY

The most crucial feature in these justifications is the careful description of the responsibility being ascribed—its limits, demands, and the consequences attendant to fulfilling and failing to fulfill it. For this description is crucial both to whether or not the responsibility is derivable from a justifiable institution and to the range of reasons which will count as excuses from the responsibility. The description will imply the relevance of sanctions under specified conditions, the relevance of the man's acceptance of the responsibility, and so on. It may turn out that when it is fully unpacked, the demands that it makes are not derivable from any justifiable 'institution' (the military code of conduct for prisoners of war is a recent example of an argument over such an issue). It may turn out that the careful description of the responsibility nicely organizes the inquiry into the existence of excuses, just as it does for the corresponding attributions. One cannot, after all, reasonably object that the man had not assented to having the responsibility when it is clear from the statement of its terms that his assent is not relevant.

No general list of excuses is useful in a discussion of procedures, for the applicability of even the obvious ones (responsible agency, acceptance of the responsibility) is dependent on the precise description of the responsibility. It may be that it is always wrong to ascribe responsibility to a man who cannot do it—does not have the requisite abilities—but this is not a matter which can be determined procedurally. One would have to show that there are no justifiable practices (institutions) which entail the ascription of responsibility without regard to agency. This may not be possible. Consider first that the ascription of responsibility does not have to carry the threat of sanction. Then recall that an effective way of developing a person's ability to do some things is just to give him the job (the responsibility) in the hope that he will develop the ability (become a responsible agent). This occurs not only as a therapy in mental illnesses, but as a standard practice in some professions—notably college teaching. So even with the matter of the responsible agency excuse, the relevance is apparently not a necessary one.

To summarize: the procedures for justifying deontic responsibilities are these: first to get a detailed description of the responsibility ascribed; next to apply the procedures developed for justifying obligations; then to justify ascribing the responsibility to the individual specified (as opposed to just anyone; of course if the responsibility can be ascribed to just anyone, no problem arises here); and finally to consider whether or not relevant excuses exist. The criteria used in the second step will again come from the description of the deontic responsibility.

XIX

AGENT MORALITY: THE CONCEPTS OF JUSTICE AND A GOOD MAN

The arguments of foregoing chapters have exhibited justification procedures for diverse types of moral judgments: value judgments, evaluations, obligation statements, attributions and ascriptions of responsibility, as well as responsible agency statements. Thus in terms of justification procedures, the major concerns of axiologists and deontologists have been dealt with, and some indication has been given as to how these two approaches might be co-ordinated.

But agent morality so far has been ignored, except for promises here and there that procedures could be developed for co-ordinating it with the others. The time has come to make good on those promises.

To do this, the concept of justice and the concept of a good man will be examined—'justice' because it exemplifies one of the leading ideals of concern to agent moralists, and 'the good man' because it is an attempt at a summation of 'moral character.' These concepts will be examined first in terms of the procedures so far developed. For it is clear that matters of justice, character assessment, and ideals cannot be separated from questions of value, obligation, and responsibility. But it will be equally clear as the discussion proceeds that axiological and deontological accounts of such matters are not totally satisfactory. So some additional procedures will be recommended.

I JUSTICE

Most discussions of justice begin with remarks on Aristotle's analysis.[1] Aristotle distinguishes a very broad sense of 'justice' from a very narrow sense. In its broad sense for Aristotle, 'justice' simply refers to the totality of traits of character which produce right conduct, and he holds that, in this sense, justice 'is not a part of virtue, but virtue entire.' He explicates this sense of justice by reference to the disposition to act in accord with what is lawful—meaning not only the laws made by governments, but 'moral law' as well. The narrower sense of justice—justice as a 'part' of virtue—is explicated initially in terms of fairness, or equity.

As usual, Aristotle's distinctions are important ones, though the way they will appear in the analysis to follow will differ somewhat from his exposition. To wit:

JUSTICE AS THE EQUITABLE

The notion of equity is undoubtedly a necessary part of the concept of justice.[2] It is a formal element of the concept, whether in matters of the making or matters of the administration of the standards of just conduct. 'Similar persons in similar circumstances are to receive similar treatment' is a rough rendition of the principle as it operates in administrative contexts. 'Standards must be written for uniform application' (i.e., must not encourage either preferential or prejudicial applications) is akin to the rendition used in setting up and criticizing standards.

A justification for acting equitably is not needed here, for the commitment to acting out reasoned conclusions about what to do cannot fail to regard all versions of the principle of equity as proceeding directly from a feature of reasoned argument: that if a set of evidence ever warrants a conclusion, then exactly the same evidence will always warrant exactly the same conclusion; a change in the conclusion requires a warranting change in the evidence. More formally, the principle of equity follows the principle of the indiscernability of identicals: $[(A = C) . (A = B)] \supset (B = C)$. Translated to the case of equity, this requires that if one ever has reasons which warrant the establishment of a certain standard for conduct, then as long as one has exactly these reasons, and none that warrant a change, that standard remains warranted. Further, when one has reasons which warrant a certain application of a standard, he has a judgment of the form: 'Whenever exactly these reasons obtain, exactly this application is warranted and no other.'

Arguments then abound about when one can be said to have 'the same' reasons. But in any case, equity is a notion which is inescapable in discussions of justice and it is in terms of equity that laws are most often criticized when their justice (or lack of it) is attacked. Discrimination, favoritism, prejudice, bias—all of these epithets charge the existence of inequities (when applied to laws, inequities in content or administration).

But *identifying* justice—even one sense of 'justice'—with the notion of equity or fairness leaves one somewhat uneasy. After all, the formal principle of equity can go only so far in ruling out content to moral judgments, and one suspects there are cases in which a principle one would regard as immoral might be applied perfectly equitably—leading to the uncomfortable conclusion that some acts that are just are not moral. Further, if one examines the way in which people criticize laws as being unjust, one will find that they not only require that laws be evenhanded to be just, but also that they not be entirely arbitrary. That is, one wants not only evenhandedness, but non-arbitrary and evenhanded principles.[3]

So while it is of course possible to sort out the 'senses' of 'justice' so that the notion of equity or fairness is regarded as 'one sense,' this may be unnecessarily confusing. The concept of equity is probably better regarded as a necessary part of the exposition of the concept of justice.

JUSTICE AS THE JUSTIFIABLE

If one is to demand, then, of the explication of 'justice' that it express an insistence that, for a law to be just, it must not only be equitable, but non-arbitrary as well, perhaps some light can be shed by considering 'the just' to be 'what is justifiable'—meaning reasoned justification as that notion has been developed in the preceding chapters.[4]

Equity would thus continue to be seen as a formal, indispensable element in the concept, operable as a criterion for or against the justness of a judgment, because it is simply an exemplification of one formal element of reasoned procedures. And what is justifiable by these procedures is certainly non-arbitrary in a strong sense of that term. On this view, then, the moral judgments obtained by reasoned procedures would *ipso facto* be just—including all the sorts of judgments dealt with by usual theories of justice—judgments to the effect that 'a principle ought to be written into law' and that '*p* ought to be administered by doing *x*' and that '*p*, administered by doing *x*, is applicable to case *a*.' Judgments of all three types can be submitted to the justification procedures so far outlined. The procedures also apply without regard to how casuistical or theoretical the judgment is, or whether it is distributive or retributive justice being considered.

This concept of justice is really very close to what Aristotle had in mind by the broad sense of justice—at least when the emphasis on the lawful is replaced by an emphasis on giving a reasoned defense of a moral judgment (whether a 'law,' rule, principle, or casuistical judgment). Justice as a virtue, then, is indeed the whole of virtue: namely, right conduct—as long as what is right is defined not *only* as what can be justified as the best of the available alternatives (in terms of value), nor *only* as what can be justified as an obligation, nor *only* as what can be justified as an act in (highly moral) character—but is defined as what is justifiable in terms of a co-ordination of all those justifications.

The just man, then (for all practical purposes synonymous with 'the good man') refers to one disposed—as a matter of character—to do what is justifiable.

2 THE GOOD MAN

A more detailed look at the notion of the just or 'good' man may be instructive, however. For one thing, good man judgments are often simply valuations. People occasionally mean to indicate the results of an affective or even summary valuation by saying that so and so is a good man. And they often mean to express the results of a calculative valuation—as when an employer, asked specifically about the job qualifications of one of his employees, says, 'He's a good man.' Such judgments, in so far as they are straightforward value expressions, can be dealt with by the procedures already outlined.[5]

Of more interest, here, however, are the cases in which something quite special is indicated by good man judgments which should probably not be taken as a straightforward value expression. The way in which good man judgments are ordinarily used in moral philosophy (and, one assumes, often used in everyday discourse) is to assert that the man referred to has a complex set of characteristics: e.g., that he is a man who is disposed to do the right thing for the right reasons. Warrant for this definition can be gotten by examining it part by part.

In the first place, a 'good man' in the usage under discussion is not merely one who tries to do (or wills to do) the right thing. Nor is he one who simply tries to do what he ought to do (is obligated to do, or is duty-bound to do, or is responsible for). He is rather one who does (or tries to do) things *for the right reasons*. 'Reasons' (motives, in this case) can be valued and evaluated. There are thus good, better, best and right motives for various pieces of conduct —motives that one 'ought' to have (at least in sense (1) of 'ought' —see Chapter X), and motives one ought not to have. The good man is one whose motives are good and right and ones he ought to have. Indeed, we are often willing to say that a man who does the *wrong* thing for the right reasons has not at all tarnished his character as a 'good man.' To be a good man is not necessarily to be adept.

But it does not seem to be true to say that the motives (or good will) are always the only thing relevant. 'The road to hell is paved with good intentions' is a cliché by which we remind ourselves that a man who consistently does the wrong things for the best of reasons is one with serious shortcomings in terms of what is expected of a 'good man.' True, the shortcomings may not be as fatal to the legitimacy of describing him as a good man as are those exhibited by the man who consistently acts from the wrong motives. But they are damaging none the less. One may, of course, occasionally 'make a mistake'—i.e., do the wrong things for the wrong reasons, or the right things for the wrong reasons— without moving people to stop calling him a 'good person.' But consistent behavior of these kinds is always, sooner or later, fatal to the description.

Just how often, in our actual practice, one may do the *wrong* things 'with the best intentions' before losing title to the description 'good man' is probably not an answerable question. It is doubtful if there is an operative general criterion for deciding such cases. It may well be possible to define and justify one, however, by applying the valuational and evaluational procedures previously discussed. The problem would be to find the right (or at least *a* right) criterion. It might be objected, of course, that a search for such a criterion is wrongheaded—as is the entire characterization of good man judgments so far—because it presupposes a separation of act from motive which is spurious. But such an objection, while it has its point in other contexts, is unnecessary here. The distinction between act and motive is made merely for convenience in dealing with justification questions. No substantive results are permitted to depend on it. So this part of the good man issue exhibits no especially difficult or unusual procedural problems.

But in addition to '(generally) doing the right things for the right reasons,' it is clear that the description 'good man' is meant to indicate a *disposition* to do this. That is, we would find something peculiar in maintaining that a man who just 'happened' to do the right things for the right reasons was a good man. If a man cannot be 'counted on' to do the right things for the right reasons—if, for example, he must struggle to make up his mind afresh each time an issue of right or wrong arises, if he is not in any sense disposed or 'ready' to do the right things—then we would be uneasy about calling him a 'good man.' One hesitates to say, 'We would refuse to call him a good man,' because ordinary usage is so various in these cases. Again, it is likely that no criterion adequate for settling all cases—or perhaps even most cases—is presently in general use. But again it is clear that criteria (totally adequate or not) can in principle be justified and grounded using procedures outlined earlier.

There is at least one further element that lies behind—but is not always explicitly expressed in—good man judgments: the good man is one who has been tested. The man who, from lack of energy or opportunity, has never been tempted to do the wrong thing, or to act from the wrong motives, and has therefore never really had to *choose* to do the right thing, is not usually described as a good man. Or, rather, if he is so described, a disclaimer is usually added to the effect that the description is not to be taken in the usual way. (Upon medical examination, his virtuous behavior was found to be the result of a glandular deficiency. . . .) But again a general principle for deciding the amount of 'testing' required is not at all clear in practice.

3 PRESUMPTIVE CRITERIA FOR ASSESSING MORAL CHARACTER

One may well object at this point, however, that while it is important to discuss procedures for justifying good man statements of the sort just explicated, and while it is possible to regard (the virtue) justice simply as the disposition to do what is justifiable, nonetheless the crucial problems concerning moral character remain untouched. That is, quite apart from what is merely 'right' to do, what sort of man exemplifies the best in human nature? What traits of character are most admirable? How is one to define excellence of character—not only in terms of what can be expected or demanded (i.e., that the man do what is right, what he ought to do)—but in terms of what it is possible to achieve beyond 'the call of duty,' beyond the merely meritorious. Right conduct is one thing, heroism is another. The disposition to do what can be justified—the disposition to do the right things for the right reasons—is one thing; nobility, heroic stature, the human excellences we stand in awe of, are quite another.

However necessary, the business of assigning responsibilities and seeing to it that they are carried out is unpleasant. However necessary, and however pervasive in our everyday affairs, the process of assessing the value of things can be a demeaning chore. And when a man does just what he ought, just what is 'the best of the available options,' there is no cause for exultation, for joy, for excitement. There is, at best, cause only for satisfaction.

It is rather the conduct that surpasses expectation, that goes beyond what can be demanded, that defines the best in human conduct in extremities in which we are prepared to excuse almost any action—it is this conduct—or rather the readiness for it—which we so often use as a standard against which to judge ourselves and others.

Yet the same problems of justification occur for these matters as for questions of value and obligation. Which of the possible ideals are the moral ones—the justifiable ones? What sort of person should I strive to be? Of course the valuation and evaluation procedures already discussed are applicable to such questions. Their relevance can hardly be denied. But just as the use of an axiological scheme to justify obligation statements may be felt to dilute some of the needed emphasis in such statements, so the reference to a 'balance of values' may be felt to be inappropriate (or at least an incomplete) answer to questions concerning ideals. Some of this dissatisfaction is no doubt simply 'feeling' —and of no importance as a reasoned objection. But it is interesting to note that the features of life heretofore used as presumptive criteria both for values and for obligations can also function as criteria for accepting or rejecting various character traits and establishing a few priorities among them.

DISTILLATES OF THE CHARACTERISTICS OF THE CRITERIA

There are some traits of character which are simply distillates of various features of the criteria: abilities of all sorts, for example, are right at the heart of what purposiveness is all about. Physical ability, intellectual ability, leadership ability, adaptability. . . the list is long. Quite obviously, the specific abilities honored in men and women vary from culture to culture, time to time, circumstance to circumstance. Not all cultures honor competitive abilities, for example, as much as ours. But I know of none which does not honor *some* ability, *some* achievements, *some* examples of excellence. The ability to achieve, and by extension, to excel, is a built-in orientation of our purposiveness. The specific forms this orientation should take—the specific abilities honored—are most reasonably decided in terms of an axiological scheme, where consideration can be given to the consequences of the various possibilities. But the cultivation of some sort of ability is simply an operative priority of the purposive nature of our lives, and as long as no reasons can be offered in opposition, this operative priority can function as a presumption in favor of the general virtue of 'being able'—and by extension, 'excelling.'

In the case of the personal nature of life, one can easily see that the virtues of integrity (in the sense of being one 'whole' thing, with respect to which some things, some actions, are just alien, unfitting, not in character—not, pardon the cliché, 'true to oneself'), as well as self-esteem, being 'self-possessed'—and even 'having a sense of responsibility' in so far as that involves the awareness of oneself as an *agent* (as opposed to patient)—are really only explications of the nature of, or extrapolations of the nature of, being a 'person.' The sense of being an integrated whole for which acts 'out of character' would be disfunctional is no more than a product of a developed sense of one's limits. Likewise, self-esteem and self-possession are built-in orientations of 'boundary-keeping' activities. The disposition toward self-preservation, if that can be said to be a virtue, is also a built-in orientation.

In terms of the aesthetic nature of life, one might argue for an orientation towards intensity in emotion, feeling, and sensation: an exuberant or passionate way of life. Also, of course, for an orientation toward those dispositions which are *productive* of the aesthetic experiences toward which we are inclined.

In any case, in so far as the personalness and aesthetic nature of our lives contain operative orientations toward some character traits, and in so far as no reasons in opposition can be found, these orientations can function as presumptions in favor of corresponding 'virtues.' This is not to say, of course, that the version of these virtues which is found appropriate will be decided in this presumptive way. Such matters will most likely be settled, if at all, by a process of evaluation.

FACILITATION OF THE CHARACTERISTICS OF THE CRITERIA

But there is another set of character traits toward which purposiveness, personalness, and our aesthetic natures are oriented: namely, those dispositions which enhance or facilitate the functions of these features of life. To say this is to say no more than that dispositions congruent with, compatible with the momentum of, or 'along the same line as' these ongoing operations meet no resistance from them. And so there is an operative orientation towards traits that are properly described as self-affirming and productive, and a resistance to traits properly described as self-destructive, debilitative, and enervating.[6]

These orientations can also function as presumptions in favor of the corresponding virtues—as long, again, as it is noted that the specific forms taken by those virtues remains a question to be decided by other means.

EQUILIBRIUM AND THE FULLNESS OBJECTION

Finally, one cannot help remarking that the very plurality—and presumptive equality—of the three features of life used as criteria, together with the fact that conflicts arise among them (e.g., over immediate, sensual gratification versus the difficult road to excellence in some regions of life) gives an operative orientation toward the maintenance of a rough equilibrium among the virtues recommended by each of the three. The sort of objection we make against a man whose character is dominated by traits developed in line with only one or two of the features illustrates this orientation.

Consider the objection against the 'emotionless' man—cold, machine-like, we say, not having a 'full' life. Of course what *counts* as 'emotionless,' what counts as an equilibrium, varies widely from one culture to another. On this score there is much room for argument in terms of the values involved. But the moralists who develop the notion of the 'harmony of the soul' as fundamental to virtue, and those who develop objections to specific moral principles as repressive of 'natural instincts' and needs, speak for the orientation toward an equilibrium. This orientation, in the absence of reasons in opposition, can also function as a presumption in favor of a corresponding virtue.

In summary, then, in addition to the arguments one can construct about moral character on the basis of axiological and deontological schemes (and the importance of such arguments must not be minimized), purposiveness, personalness, and the aesthetic nature of our lives contain operative orientations which can be used as presumptions in favor of corresponding virtues: virtues which are simply distillates of the nature of the three features of life mentioned, virtues which enhance the functioning of those features, and virtues productive of an equilibrium among them. Agent morality considerations thus have a starting point outside of axiological and deontological schemes.

NOTES

1 For samples of analyses beginning with helpful comments on Aristotle, see Nicholas Rescher's *Distributive Justice* (Indianapolis: Bobbs-Merrill, 1966) and Chaim Perelman's *Justice* (New York: Random House, 1967).

2 See John Rawls, 'Justice as Fairness,' *Philosophical Review*, 67 (1958), pp. 164–94, and 'The Sense of Justice,' *Philosophical Review*, 72 (1963), pp. 281–305. There are criticisms by E. W. Hall ('Justice as Fairness: A Modernized Version of the Social Contract,' *Journal of Philosophy*, 54 (1957), pp. 662–70) and by Robert Paul Wolff ('A Refutation of Rawls's Theorem on Justice,' *Journal of Philosophy*, 63 (1966), pp. 179–90). The issues here have become as tangled as most in moral philosophy.

3 See Rescher, *Distributive Justice* (Indianapolis: Bobbs-Merrill, 1966); Henry Sidgwick, Book III, Chapter V, of *Methods of Ethics* (reissue of seventh edn, 1907, Chicago: University of Chicago Press, 1962); and for a very illuminating discussion of the whole of the issue, Perelman, *Justice* (New York: Random House, 1967).

4 Both natural law theories of justice and explications of justice as respect for persons will find some elements of their views (although no doubt not all, and not with precisely the emphasis or interpretation they use) submerged in the procedures outlined in these pages. In particular, natural law theories will find some limited kinship with the presumptive criteria outlined below, and at least one of the criteria—personalness—is closely tied to the 'respect for persons' view. See, for this latter view, W. G. McClagan, 'Respect for Persons as a Moral Principle,' *Philosophy*, 35 (1960), pp. 193–217 and 289–305.

5 The sense of 'good man' explicated as 'a man disposed to do the right things for the right reasons' may be treated as the expression of a good-of-its-kind valuation. Indeed, one often gets the feeling, in reading the Greeks on virtue, for example, that they mean to say precisely that the good man is good *as* a man—that the qualities he exhibits define what it means to excel as a human being, to be an excellent example of the species. If good man judgments are treated in this way, perhaps the route to justification will look a bit more like axiology than 'agent morality.'

6 To say there are operative priorities in favor of self-affirming and productive traits is *not* to say that these priorities are always followed. Circumstances, and the cumulative effects of adaptations to them, often result in traits that are quite self-negating, self-destructive. It is only to say that if one expects to *justify* various character traits, he gets this sort of presumptive help for the productive ones, but not for the destructive ones. The openness of the procedure, however (the merely *presumptive* validity of these orientations), leaves open the possibility that contrasting traits might be justified.

XX

CONCLUDING REMARKS

The analysis of justification procedures proposed at the outset is now complete, and a few remarks about what it has accomplished are in order. Has the plausibility of scepticism been destroyed? Even if the procedural arguments made here are correct, is it clear that these procedures are usefully applicable to significant cases? If they are not, then the sceptic's claims are as viable as ever.

Whether the arguments here succeed or not is a matter to be decided in the dialectic of further philosophical inquiry. But if they are correct, then it seems clear that they can be useful in concrete situations. They are not, of course, *decision procedures*—in the technical sense of procedures, which, when applied by some clerical, mechanical routine, always yield decisions. That fact has been acknowledged time and again in these pages with respect to the inclusion of a *ceteris paribus* clause in nearly all of the steps of the procedure. But it has also been maintained that dealing with the *ceteris paribus* clauses is often less difficult in practice than in theory—and thus that the procedures can *occasionally* yield decisions, even if they are not decision procedures. That fact, when combined with the fact that it is the broadest, most consequential matters (e.g., questions of homicide) which seem most amenable to the procedures, seems enough to meet the challenge of scepticism.

Still, there may be some discontent with two other broad features of the analysis: the choice of presumptive criteria; and the synthesis of axiology, deontology, and agent morality.

As to the choice of criteria: purposiveness, personalness, and the aesthetic nature of life may seem uninspired at the very least. They are certainly superficial as characterizations of 'human nature'—especially to generations nourished on the sophisticated speculations of post-Freudian personality theory. But it was not part of the plan to communicate a sense of the grandeur and depths of human nature. The point was simply to illustrate grounding procedures—*with examples as unobjectionable as possible.* An unobjectionable example is, in one sense, inevitably a trite one. But an objectionable example, however profound and moving, invites distraction from the main line of argument. If the procedure is sound, then concern can reasonably turn to a more satisfactory analysis of actual (as opposed to illustrative) criteria.

As for the co-ordination of axiological, deontological, and agent morality approaches proposed in these pages, it may be felt that what has really been proposed is a *synthesis* of the three approaches into a fourth—a kind of grandiose symphony of traditional themes. If so, some disclaimers are needed.

No implication was intended here that a grand harmony of the results of axiology, deontology, and agent morality was possible or even desirable. The point was only that each of the approaches has legitimacy as a co-ordinate part of moral theorizing. Each has a unique contribution to make, and judgments of each type have a degree of independent access to justification. Arguments of each sort have independent starting points. There is thus no problem of 'logical priority.'

But it is apparent also that questions from all three points of view will be relevant to a rational assessment of any significant moral situation. It is rarely adequate to analyze situations involving violence, for example, solely in terms of the values involved, solely in terms of duty, obligation, and responsibility, or solely in terms of the character of those involved. Indeed, it is doubtful that one can ever get an adequate analysis of such situations—at least in terms of one that smoothes out all the sources of moral concern. But it is clear that attempts to subordinate one sort of moral approach to another merely complicate the problem of getting adequate analyses of moral situations. So what is argued for here is the abandonment of interscholastic *competition*, not the papering over of genuine differences and genuine conflicts.

In summary, then, what has been accomplished in these pages—

if the arguments in them are substantially valid—is no more and no less than what can be expected of a prolegomenon to moral theory: a demonstration that a rational moral theory is both possible and worthwhile, and a schema for beginning its construction. To demand more is to demand a substantive moral theory. If the arguments of this book are correct, that step has been made substantially easier.

INDEX

International Library of Philosophy & Scientific Method

Editor: Ted Honderich

(Demy 8vo)

Allen, R. E. (Ed.), **Studies in Plato's Metaphysics** *464 pp. 1965.*
Plato's 'Euthyphro' and the Earlier Theory of Forms *184 pp. 1970.*
Allen, R. E. and Furley, David J. (Eds.), **Studies in Presocratic Philosophy** *326 pp. 1970.*
Armstrong, D.M., **Perception and the Physical World** *208 pp. 1961.*
A Materialist Theory of the Mind *376 pp. 1967.*
Bambrough, Renford (Ed.), **New Essays on Plato and Aristotle** *184 pp. 1965.*
Barry, Brian, **Political Argument** *382 pp. 1965.*
Bird, Graham, **Kant's Theory of Knowledge** *220 pp. 1962.*
Bogen, James, **Wittgenstein's Philosophy of Language** *256 pp. 1972.*
Broad, C. D., **Lectures on Psychical Research** *461 pp. 1962. (2nd Impression 1966.)*
Crombie, I. M., **An Examination of Plato's Doctrine**
I. Plato on Man and Society *408 pp. 1962.*
II. Plato on Knowledge and Reality *583 pp. 1963.*
Day, John Patrick, **Inductive Probability** *352 pp. 1961.*
Dennett, D. C., **Content and Consciousness** *202 pp. 1969.*
Dretske, Fred I., **Seeing and Knowing** *270 pp. 1969.*
Ducasse, C. J., **Truth, Knowledge and Causation** *263 pp. 1969.*
Edel, Abraham, **Method in Ethical Theory** *379 pp. 1963.*
Farm, K. T. (Ed.), **Symposium on J. L. Austin** *512 pp. 1969.*
Flew, Anthony, **Hume's Philosophy of Belief** *296 pp. 1961.*
Fogelin, Robert J., **Evidence and Meaning** *200 pp. 1967.*
Franklin, R., **Freewill and Determinism** *353 pp. 1968.*
Gale, Richard, **The Language of Time** *256 pp. 1967.*
Glover, Jonathan, **Responsibility** *212 pp. 1970.*
Goldman, Lucien, **The Hidden God** *424 pp. 1964.*
Hamlyn, D. W., **Sensation and Perception** *222 pp. 1961. (3rd Impression 1967.)*
Husserl, Edmund, **Logical Investigations** *Vol. I: 456 pp. Vol. II: 464 pp.*
Kemp, J., **Reason, Action and Morality** *216 pp. 1964.*
Körner, Stephan, **Experience and Theory** *272 pp. 1966.*
Lazerowitz, Morris, **Studies in Metaphilosophy** *276 pp. 1964.*
Linsky, Leonard, **Referring** *152 pp. 1967.*
MacIntosh, J. J. and Coval, S. C. (Eds.), **Business of Reason** *280 pp. 1969.*
Meiland, Jack W., **Talking About Particulars** *192 pp. 1970.*
Merleau-Ponty, M., **Phenomenology of Perception** *487 pp. 1962.*
Naess, Arne, **Scepticism** *176 pp. 1969.*
Perelman, Chaim, **The Idea of Justice and the Problem of Argument** *224 pp. 1963.*
Ross, Alf, **Directives, Norms and their Logic** *192 pp. 1967.*
Schlesinger, G., **Method in the Physical Sciences** *148 pp. 1963.*
Sellars, W. F., **Science and Metaphysics** *248 pp. 1968.*
Science, Perception and Reality *374 pp. 1963.*
Shwayder, D. S., **The Stratification of Behaviour** *428 pp. 1965.*